ELECTROSTATIC BASIS OF BIOLOGICAL ACTIONS

Other World Scientific Titles by the Author

From Kibbutz Fishponds to The Nobel Prize: Taking Molecular Functions into Cyberspace
ISBN: 978-981-12-4178-9
ISBN: 978-981-12-4315-8 (pbk)

ELECTROSTATIC BASIS OF BIOLOGICAL ACTIONS

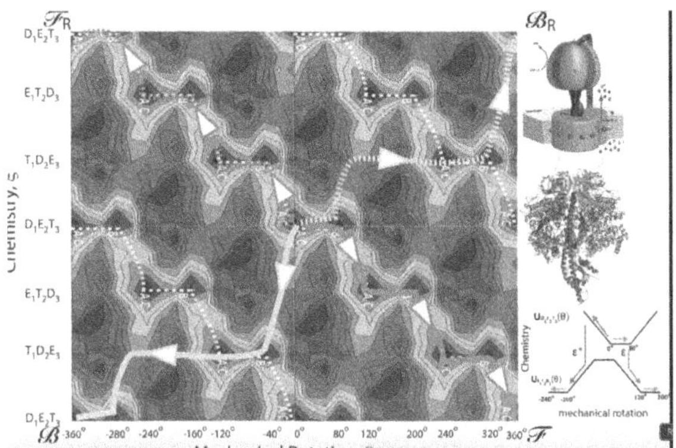

Arieh Warshel
University of Southern California, USA

World Scientific
NEW JERSEY • LONDON • SINGAPORE • BEIJING • SHANGHAI • HONG KONG • TAIPEI • CHENNAI • TOKYO

Published by

World Scientific Publishing Co. Pte. Ltd.
5 Toh Tuck Link, Singapore 596224
USA office: 27 Warren Street, Suite 401-402, Hackensack, NJ 07601
UK office: 57 Shelton Street, Covent Garden, London WC2H 9HE

Library of Congress Control Number: 2025032359

British Library Cataloguing-in-Publication Data
A catalogue record for this book is available from the British Library.

ELECTROSTATIC BASIS OF BIOLOGICAL ACTIONS

ISBN 978-981-98-2060-3 (hardcover)
ISBN 978-981-98-2061-0 (ebook for institutions)
ISBN 978-981-98-2062-7 (ebook for individuals)

For any available supplementary material, please visit
https://www.worldscientific.com/worldscibooks/10.1142/14507#t=suppl

Desk Editor: Shaun Tan Yi Jie

Typeset by Stallion Press
Email: enquiries@stallionpress.com

Acknowledgments

I would like to thank my wife Tamar for her constant support during the last 60 years and of course during the time of writing this book. I also thank my daughters Merav Efrat and Yael Warshel and their families. I am grateful to Professor William Parson who contributed in a major way to my key works and to our 2006 BBA review that was used extensively in the book. I would also like to thank my coworkers who contributed to many of the works outlined in the book, and to Andrew Levy who helped in editing the book.

Preface

Biological systems control all life processes, and understanding how they function is essential for both medical progress and fundamental scientific advances. Gaining such understanding requires the ability to infer key functional properties from structural — and even sequence — information. Therefore, identifying the physical factors that govern biological function is of central importance. This book presents, step by step, the evidence supporting the view that electrostatic effects are the most significant factor linking the structure and function of biological molecules. The corresponding discussion also involves descriptions of the action of important biological systems.

The book begins with a brief overview of molecular forces and simulation methods, followed by a detailed exploration of electrostatic theories. This discussion emphasizes the microscopic foundations of electrostatic modeling in biological systems and highlights its crucial role in moving beyond the limitations of macroscopic approaches that rely on bulk dielectric constants. The book then provides a comprehensive review of a wide range of biological functions. In each case, we examine simulation studies demonstrating that the most accurate description of the observed function emerges from analyzing the underlying electrostatic energetics.

As background for the perspective of this book, it may be helpful to recount my path to the realization that electrostatic effects play a major role in biology. In 1964, during my third year at the Technion, I joined a research project under the guidance of Professor Yechiel Shalitin. In this project, I investigated the effect of ionic strength on the catalytic reaction of chymotrypsin. Since I observed only a very small effect, we concluded — incorrectly — that electrostatic interactions did not contribute significantly to enzyme catalysis. This conclusion was later published as an abstract in a conference of the Israel Chemical Society[1].

Subsequently, in my PhD project (1966–1969) which led to the development of the consistent force field[2], we paid little attention to electrostatic effects, as the refinement process involved mainly hydrocarbons. As a result, I remained unaware of the importance of electrostatics in biological systems. Even in 1973, when I co-authored a paper with Neville Kallenbach on the spectral properties of cytosine base pairs, we did not explore the solvent effect on the stacking process. Although the paper was accepted for publication in *Biopolymers*, I chose to delay its publication in 1974 due to concerns over this omission. It was only in 1979 that I published the portion addressing the solvent effect on stacking[3]. A major turning point came in 1974, during my time at the Medical Research Council in Cambridge, England. It was there that I first heard Max Perutz express his belief that electrostatic interactions play a crucial role in protein function. However, my true appreciation of electrostatics began when I started modeling the catalytic reaction of lysozyme with Mike Levitt. This project forced me to realize the enormous role of the solvent on the reaction energy. Specifically, I found that the calculated energy required to break a carbon-oxygen bond in my enzyme model exceeded the experimental value in solution by more than 40 kcal/mol. This large discrepancy led me to realize that I had neglected the massive contribution of solvation energy. This realization compelled me to develop a practical model for solvation by water and/or protein. At this time, I found only preliminary studies on solvation using continuum dielectric models[4–8] or a few isolated water molecules (e.g., Ref. 9). However, I could not base my future attempt to pinpoint the origin of enzyme catalysis on the undefined cavity concept in the continuum model, nor on using a few water molecules instead of the complete enzyme-plus-water environment. In fact, these approaches offered little guidance on how to treat electrostatics within proteins meaningfully. Faced with these limitations, I delved into the literature on electrostatic theory and gradually came to understand that macroscopic electrostatics — and its poorly defined dielectric constants — were inadequate. What was needed was a microscopic model grounded in clear physical principles. Given the available computing power at the time, I decided to develop a simplified dipolar model for water and a polarizable force field, with iteratively updated induced dipoles, for proteins[10]. This allowed us to eliminate reliance on macroscopic concepts, including the ambiguous notion of a dielectric constant, and to replace them with a microscopic model based on physically meaningful interactions. Our studies on lysozyme, and later on other systems, gradually led me to the conclusion that electrostatic interactions are the key factor in enzyme catalysis.

The idea of electrostatic catalysis was not widely accepted, and at times it was tempting for some to dismiss it as the proposal of a "wild swashbuckler" scientist. An interesting example is a 1991 article by Anne Moffat on the field of molecular modeling, published in the National Science Foundation magazine *Mosaic*. The article stated that I believed all of the catalytic power of enzymes was associated with electrostatic effects, while Karplus and McCammon, taking a more "Catholic approach" that included molecular dynamics and free energy calculations, concluded that other effects could also be important. The piece then shifted to a discussion of QM/MM calculations of enzymes. The glaring problem was that, at the time of the interview, I was the only researcher consistently using molecular dynamics and free energy calculations, as well as QM/MM modeling, in realistic studies of enzyme catalysis. These methods formed the basis of my conclusions about the importance of electrostatic effects — conclusions I reached only after systematically evaluating all other relevant possibilities. It seems to me that Moffat was not negatively inclined, but simply could not believe that my assertions could be correct, given the apparent simplicity of my final conclusion about the role of electrostatics.

Over time, I came to realize that electrostatic effects may be the key to many biological functions, ranging from energy transduction to signal transduction and beyond. Other researchers began to reach similar conclusions based on macroscopic considerations, but resistance to the idea remained strong. For example, my first paper on electron transfer in cytochrome c was rejected by the *Journal of Molecular Biology*, where a distinguished referee objected to the claim that the process is controlled by electrostatic (solvation) energy — overlooking the fact that we were dealing with a charge transfer process. Similarly, my early work on proton transport was largely ignored until around 2004, at a time when much of the scientific community was focused on the Grotthuss mechanism, in which proton motion is assumed to be almost barrierless. This view overlooked the significant solvation barriers involved in moving protons between different sites within proteins. Resistance to the idea that electrostatic effect plays a major role in enzyme catalysis has been very strong, considering all the competing proposals that were favored by the biochemical community. Nonetheless, it is now clear that electrostatic effects play a major role in determining biological function — and that they serve as major descriptors in the correlation between structure and function.

Overall, the book provides both a historical perspective and a general overview of computational methods for evaluating electrostatic interactions in macromolecular systems, as well as the role of electrostatic effects in biological molecules. The book can serve as a guide for simulating biological functions, with an emphasis on electrostatic aspects. Additionally, it offers a review of the behavior of key biological systems and the insights that have emerged from simulation studies.

About the Author

Dr. Arieh Warshel is Distinguished Professor of Chemistry and Biology at the University of Southern California, USA, where he currently holds the Dana and David Dornsife Chair in Chemistry. He was awarded the 2013 Nobel Prize in Chemistry, alongside Martin Karplus and Michael Levitt, for the development of multiscale models for complex chemical systems. He is an elected Member of the USA National Academy of Sciences, a Foreign Member of the Russian Academy of Sciences, Honorary Fellow of the Royal Society of Chemistry, and elected Fellow of the Biophysical Society and the American Association for the Advancement of Science. He has authored over 500 peer-reviewed articles and two books, *Computer Modeling of Chemical Reactions in Enzymes and Solutions* (Wiley, 1991) and *From Kibbutz Fishponds to The Nobel Prize* (World Scientific, 2021).

Dr. Warshel was born in 1940 to a family of founders of the Sde Nahum kibbutz, in what was then Mandatory Palestine (now Israel). He served in the Israeli Armored Corps, and fought in both the 1967 Six-Day War and the 1973 Yom Kippur War, attaining the final rank of Captain. He attended the Technion in Haifa, where he studied Chemistry based on an off-the-cuff suggestion by a friend. He received his BSc degree, summa cum laude, in 1966 — the same year he married his wife Tamar, with whom he has two daughters. He earned his MSc (1967) and PhD (1969) in Chemical Physics at the Weizmann Institute of Science, working under the Institute's scientific director, Shneior Lifson, and collaborating with Michael Levitt. After his PhD, he did postdoctoral work at Harvard University with Martin Karplus from 1970–1972. He then returned to the Weizmann Institute and also worked at the Laboratory of Molecular Biology in Cambridge, England where he was reunited with Michael Levitt. He joined the Chemistry faculty at USC in 1976, where he has been ever since.

Dr. Warshel is known for his work on computational biochemistry and biophysics, in particular for pioneering multiscale simulations of the functions of biological systems, and for developing what is known today as Computational Enzymology. He is responsible for introducing molecular dynamics in biology; developing the QM/MM approach; introducing simulations of enzymatic reactions; pioneering microscopic simulations of electron transfer and proton transfer in solutions and in proteins; pioneering microscopic modeling of electrostatic effects in macromolecules; and co-introducing simulations of protein folding. He has continued to be active in research, and in April 2017 he opened the Warshel Institute for Computational Biology at the CUHK Shenzhen campus, with the intention of fostering one of the world's most advanced computational biology centers in the Southern Chinese city. As of 2025, he has a h-index of 122.

Contents

Chapter I

Fundamentals of Molecular Modeling

This opening section provides an introduction to different approaches that allow one to model molecular systems. Such approaches range from quantum mechanical descriptions that can provide (in principle) accurate depictions of molecular energetics to more empirical approaches that provide practical ways for modeling large molecules. We then consider different methods that allow one to use molecular potential surfaces in simulating various molecular properties. Introducing these approaches is helpful for comprehending the material in the following chapters. However, a detailed understanding of the formulations presented here is not required to follow the subsequent discussion of electrostatic effects.

I.1 Quantum Mechanical Basis of Molecular Forces

All molecules are built from atoms which are bonded to each other by intermolecular forces that are due to the interaction between the valence electrons and the effect of the screened nuclear cores. The strength of the resulting chemical forces is determined by the quantum mechanical wave function of the system of atoms that constitute the given molecule. The quantum mechanical wave function and the corresponding energy surface can be determined by solving the Schrödinger equation[11].

$$\hat{H}\Psi = E\Psi \tag{I.1}$$

Here, $\hat{H}$ is the Hamiltonian operator that represents the kinetic and potential energy of the system. Ψ is the wave function of the electrons of the system for a fixed nuclear configuration, with the square of this function describing the probability of finding the electrons of the system in different regions of space (e.g., Fig. I.1). E is the total energy of the system. In most chemical problems, the equation can be solved within the Born-Oppenheimer approximation, where E and Ψ are evaluated for fixed nuclear distances.

The relevant solution can be obtained numerically by a wide range of approaches that have been developed in the last hundred years and have been advanced in major ways in the last 30 years[12,13], due in part

Quantum Mechanics

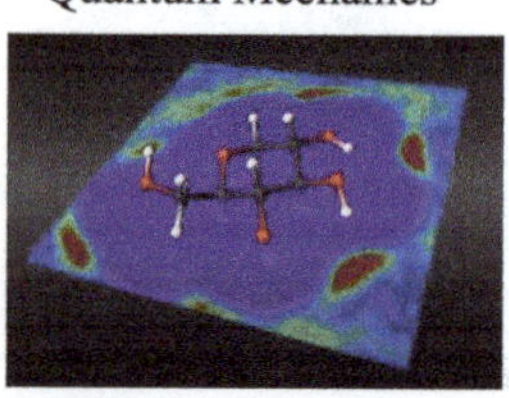

very expensive

Fig. I.1 Quantum mechanical description of a molecular system. The figure describes the electron density in a medium-sized molecule.

to the enormous increase in computer power. The dependence of the energy on the position of the atoms in the molecules is called the *molecular potential surface*. The reliability of such surfaces depends on the size of the molecule, going from very reliable surfaces for small molecules to more qualitative surfaces for large molecules.

I.2 Force Fields

Given the high computational cost of quantum mechanical calculations and the need for rapid evaluation of molecular forces, it is important to develop approaches that can provide approximate force estimates.

In cases involving small changes in bond lengths and bond angles, it is possible to obtain a very good approximation by using what is known as a *force field* (see schematic description in Fig. I.2). A force field is typically a function of the bond lengths b, the bond angles θ, torsional angles ϕ, and non-bonded distances r. Accordingly, we use a potential of the form:

$$U = \frac{1}{2}\sum_{i} K_{b,i}\left(b_{i} - b_{0,i}\right)^{2} + \frac{1}{2}\sum_{i} K_{\theta,i}\left(\theta_{i} - \theta_{0,i}\right)^{2} + U_{\phi}(\phi) \qquad (I.2)$$

The first two terms represent the potential around a very deep potential well, and because the molecule stays in most cases inside this well (except in extreme cases of bond dissociation), it is reasonable to approximate this part of the potential surface by its quadratic expansion, which is used above.

The torsional potential is a periodic function, which can be described by the leading terms in the Fourier expansion of the potential:

$$U_{\phi}(\phi) = \frac{1}{2}\sum_{i} K_{\phi,i}\left(1 - \cos n\phi_{i}\right) \qquad (I.3)$$

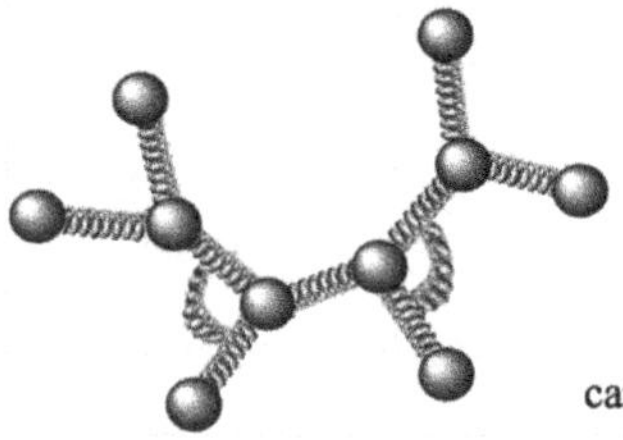

Fig. I.2 Force field description of a molecule.

The non-bonded potential can be described by an atom-atom interaction potential of the form:

$$U_{nb} = \sum_{ij} \left(A_{ij} r_{ij}^{12} - B_{ij} r_{ij}^{6} + C \frac{q_i q_j}{r_{ij}} \right) + U_{\text{ind}}(\mathbf{r}) \tag{I.4}$$

where r_{ij} is the distance between the indicated atoms, the q_i are the residual atomic charges, and U_{ind} is the many-body inductive effect of the electronic polarizabilities.

The introduction of potential functions opened the way for the use of computers in conformational analysis of medium-sized molecules (e.g., Refs. 2,14,15) and proteins[16].

I.3 Sampling Potential Surfaces

Regardless of the way the potential surface is evaluated, we have to be able to sample these surfaces effectively. This requirement starts with the need to determine the minima on the surface, which correspond to the stable structures, and extends to determining the barriers between the minima. In condensed phases, the system can adopt countless configurations that have similar energies. Fortunately, in many cases, the change in energy between physically relevant states is similar in different configurations of the system. In considering the simplest aspects of configurational sampling, we will start with energy minimization as described below.

I.4 Energy Minimization

The potential surface for a molecule with n atoms can be expanded formally around the equilibrium configuration r_0 to give

$$U(\mathbf{r}_0+\delta\mathbf{r})=U(r_0)+\sum_{i\alpha}(\partial U/\partial r_{i\alpha})\delta r_{i\alpha}$$
$$+\frac{1}{2}\sum_{i\alpha,j\beta}(\partial^2 U/\partial r_{i\alpha}\,\partial r_{j\beta})\delta r_{i\alpha}\delta r_{j\beta}+\cdots \tag{I.5}$$

where the indices i and j designate atoms while α and β run over the x, y, and z coordinates of each atom. The first term in Eq. (I.5) is just the energy of the system at equilibrium. The second term represents the deviation from equilibrium, and the $3n$ set of equations (for $i = 1, 2, \ldots, n$ and $\alpha = x, y, z$) represents the condition that r_0 is an equilibrium configuration.

$$\partial U/\partial r_{i\alpha}=0 \tag{I.6}$$

This set of equations can be solved approximately by the steepest descent method, where the nth step toward the minimum is obtained by

$$r_{i\alpha}^{n}=r_{i\alpha}^{n-1}-g\frac{(\partial U/\partial r_{i\alpha})}{\left|\sum_{i\beta}(\partial U/\partial r_{i\beta})^2\right|^{1/2}} \tag{I.7}$$

Here, g is a scaling factor which is changed in an iterative way to prevent r from overshooting; the minimum g is increased in every iteration as long as $U(r_n) < U(r_{n-1})$ and decreases if $U(r_n) > U(r_{n-1})$.

The steepest-descent method converges very slowly and is not very effective in searching for minima. A much more reliable and efficient approach is the modified Newton-Raphson method. This method is based on expanding the gradient as a Taylor series around the given $\mathbf{r}$ and finding the $\delta\mathbf{r}$ that leads to $\mathbf{r}_0$ where the gradient is zero (i.e., $\mathbf{r}_0 = \mathbf{r} + \delta\mathbf{r}$). This gives

$$\partial U(\mathbf{r}_0)/\partial r_{i\alpha} = \partial U(\mathbf{r}_0 + \delta\mathbf{r}) \cdot \partial r_{i\alpha} + \sum_{j\beta} F_{i\alpha,j\beta}\, \delta r_{j\beta} = 0 \qquad \text{(I.8)}$$

where F is the matrix of second derivatives, i.e., $F_{i\alpha,j\beta} = \partial^2 U/\partial r_{i\alpha}\partial r_{j\beta}$.

Now, solving Eq. (I.8), one obtains

$$\mathbf{r}_0 = \mathbf{r} + \delta\mathbf{r} = \mathbf{r} - \mathbf{F}^{+}\nabla U(r) \qquad \text{(I.9)}$$

where $\nabla U(r)$ is the gradient vector [with $(\nabla U)_{i\alpha} = \partial U/\partial r_{i\alpha}$] and $\mathbf{F}^{+}$ is the generalized inverse of F, which is constructed by "filtering" the zero eigenvalues of F before inverting this matrix. This is done by diagonalizing F with $\Lambda = \mathbf{S'FS}$, constructing the matrix $\bar{\Lambda}$ with $\bar{\Lambda}_{ii} = \Lambda_{ii}$ for $\Lambda_{ii} \neq 0$ and $\bar{\Lambda}_{ii} = \infty$ for $\Lambda_{ii} = 0$ and then obtaining $\mathbf{F}^{+}$ by $\mathbf{F}^{+} = \mathbf{S}(\bar{\Lambda}^{-1})\mathbf{S}'$. The reason for introducing this special inversion procedure is that $U(r)$ is invariant to rigid rotation and translation of the molecule so that the F matrix has zero eigenvectors which prevent its regular inversion.

It is instructive to note that both the steepest-descent and the Newton-Raphson methods lead in the direction of $-\nabla U$; however, the steepest-descent method is unable to tell us how far to go in each step and

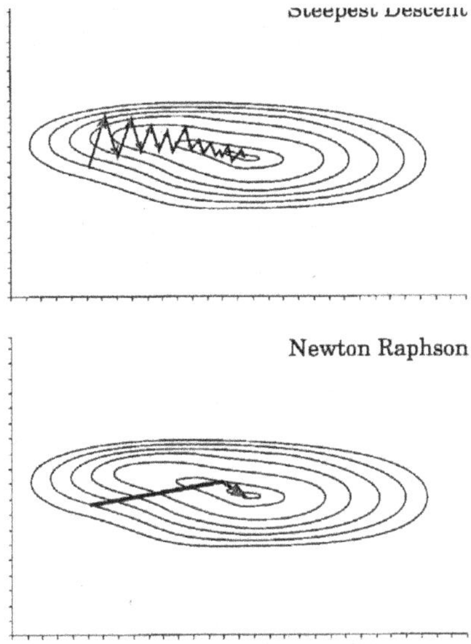

Fig. I.3 Illustrating the effectiveness of different minimization schemes. The steepest-descent method requires many steps to reach the minimum, while the Newton-Raphson method locates the minimum in a few steps (at the expense, however, of evaluating the second derivative matrix). Reproduced with permission[18].

therefore we have to search for the minimum in a very ineffective way (see Fig. I.3).

Equation (I.9) requires the evaluation of the second derivative matrix F, which is quite involved. Alternatively, one can use the *conjugated gradient* methods where an approximation of $\mathbf{F}^+$ is being built while searching the minimum using only the first derivatives vector $-\nabla U$ (for a description of these powerful **Davidon–Fletcher–Powell** methods and related approaches, see Ref. 17).

Energy minimization methods that use information about the second derivative of the potential are quite effective in the structural refinement of proteins. That is, in the process of X-ray structural determination one sometimes obtains bad steric interactions that can easily be relaxed by a small number of energy minimization cycles. The type of relaxation that can be obtained by energy minimization procedures is illustrated in Fig. I.4.

Although the conjugated gradient and related methods are very effective in finding *local minima,* they do not overcome the problems associated with the enormous dimensionality of macromolecules. That is, in systems with many degrees of freedom, we expect to find a very large number of local minima, and it is not clear how to find the lowest minimum in an efficient way. For this purpose, one must use much more computer time and different types of search procedures (**Monte Carlo** or **molecular dynamics**), generating different configurations at different regions of conformation space and locating the minimum energy structure of each region. However, such methods were largely impractical until the emergence of supercomputers in the early 1980s, and even today they do not allow for a complete search of a protein's conformational space.

I.5 Molecular Dynamics and Monte Carlo Simulations

Molecular dynamics simulations are approaches that evaluate the motion of the atoms in a given system and provide the positions or trajectory of these atoms as a function of time. The trajectories are evaluated by solving the classical equations of motion for the molecule under consideration. This is not unlike the well-known approach by which one evaluates the speed and position of a projectile, starting from the initial velocity, the mass, and the forces, by using Newton's

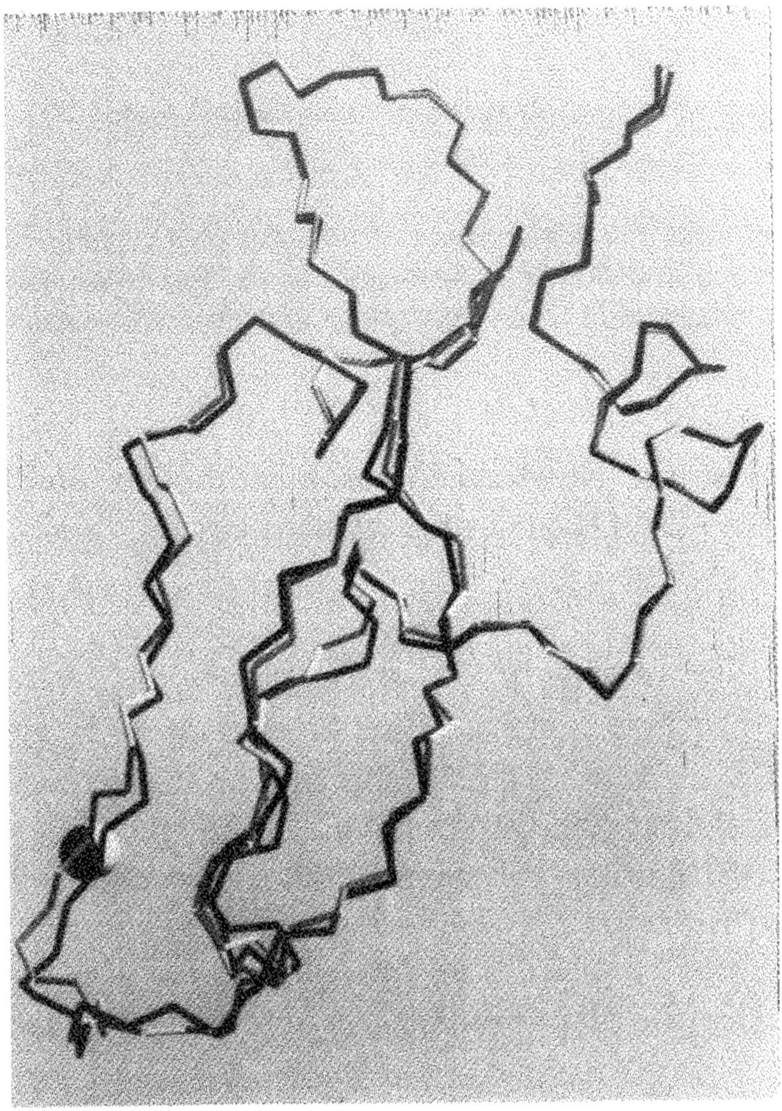

Fig. I.4 Relaxation of steric interactions by energy minimization. The figure shows the result of energy minimization of the Gly 12 → Ala mutant of the bovine pancreatic trypsin inhibitor from the native geometry (light), where it has bad steric contacts, to a relaxed geometry (dark). The mutant Alanine residue is drawn as a sphere. Reproduced with permission[18].

equation of motion. However, in the case of molecules, one obtains the relevant forces on each atom from the first derivatives of the given potential functions. The actual evaluation of classical trajectories is done numerically, expressing the changes in coordinates and velocities at a time increment, Δ*t*, using:

$$r_i(t+\Delta t) = r_i + \left[2\ddot{r}(t+\Delta t) + 5\ddot{r}_i(t) - \ddot{r}_i[t-\Delta t]\right]\Delta t/6$$
$$r_i(t+\Delta t) = r_i(t) + \dot{r}_i\Delta t$$
$$\dot{r}_i(t+\Delta t) = \dot{r}_i(t) + \ddot{r}_i\Delta t = \dot{r}_i(t) - \left[(\partial U/\partial r_i)m_i^{-1}\right]\Delta t \qquad \text{(I.10)}$$

where the dot designates a time derivative, and we use Newton's law:

$$m_i\ddot{r}_i = -\partial U/\partial r_i \qquad \text{(I.11)}$$

Starting with a given set of initial conditions [e.g., with the values of r_i $(t = 0)$ and $\dot{r}_i$ $(t = 0)$], we can evaluate $r(t)$ either by numerically integrating Eq. (I.10) or by using the somewhat more complicated, but far better, approximation:

$$r_i(t+\Delta t) = r_i(t) + \dot{r}_i\Delta t + \left[4\ddot{r}i(t) - \ddot{r}_i(t-\Delta t)\right]\Delta t^2/6$$
$$\dot{r}_i(t+\Delta t) = \dot{r}_i + \left[2\ddot{r}(t+\Delta t) + 5\ddot{r}_i(t) - \ddot{r}_i[t-\Delta t]\right]\Delta t/6 \qquad \text{(I.12)}$$

In principle, it is possible to determine averages over the potential surface without actually sampling the entire conformational space by exploiting the fact that points with high energy do not contribute significantly to the overall average. A powerful and systematic approach of doing so is the Monte Carlo method devised by Metropolis *et al.*[19] This approach generates random conformations by moving several atoms or torsional angles and uses the corresponding value of the potential surface to accept or reject each new conformation. If the energy of the new conformation (n) is lower than the previous (n–1), the new conformation is accepted. If the energy of the new conformation is higher than that of the previous, a random number, R_n, between 0 and 1 is generated and compared to a probability factor P_n:

$$P_n = \exp[-(u_n - u_{n-1})/k_B] \tag{I.13}$$

If the random number is larger than P_n, the move is rejected. If R_n is smaller than P_n, the conformation is accepted. The desired average is then simply evaluated by

$$\langle A \rangle = \frac{1}{N} \sum_{i=1}^{N} A(\mathbf{r}^N) \tag{I.14}$$

This random procedure generates a probability that corresponds to the proper Boltzmann distribution of the entire system if enough points are evaluated. In principle, one needs an enormous number of points, but in practice, one can obtain reasonable results with a reasonable amount of computer time (if the property has similar values at different conformations). In fact, Monte Carlo approaches do not involve complete random searches, because they are usually implemented with umbrella sampling or free energy perturbation approaches (see below) that bias the search toward important regions in the conformational space. Unfortunately, the Monte Carlo method is extremely inefficient.

I.6 Describing Chemical Processes by Quantum Mechanics/Molecular Mechanics Approaches

The problem with force field approaches arises when one tries to explore chemical processes that involve, for example, bond breaking and bond making. In such cases, the representation of bonds and angles by harmonic springs become invalid, and it is crucial to use quantum mechanical treatments to explore the potential energy surface. Unfortunately, evaluating quantum mechanical energy surfaces is computationally expensive and becomes significantly more so as the number of atoms increases. This situation necessitates the use of specialized methods.

The method of choice in such cases has been the so-called **hybrid quantum mechanics/molecular mechanics (QM/MM)** method[10]. This approach (described schematically in Fig. I.5) divides a large system into two regions: an *active region*, which is treated quantum mechanically, and a *surrounding region*, which encompasses the rest of the system and is treated using a classical force field.

$$U = U_{\text{classical}} + U_{\text{quantum}} + U_{\text{quantum/classical}} \tag{I.15}$$

Arguably, the key aspect of the QM/MM treatment is the coupling between the quantum and molecular mechanics parts. This coupling involves mainly electrostatic coupling, which is obtained by adding the electrostatic potential from the environment to the Hamiltonian elements that represent the ionization energies of the active atoms using:

$$H_{ii} = H_{ii,0} - V_i \tag{I.16}$$

where H_{ii} is the diagonal element of the Hückel Hamiltonian, $H_{ii,0}$ is the value of H_{ii} in the gas phase (which is basically similar to the ionization energy of the given atom), and V_i is the electrostatic potential from the surrounding environment on the ith atom. While this treatment is extremely simple, it effectively captures most of the environmental effects. In particular, the electrooptic contribution of V_i captures the most important factor in enzyme catalysis.

The specific quantum mechanical method used depends on the available computational resources and the level of accuracy required. In this respect, it is important to note that obtaining reliable results requires proper sampling over the configurations of the system (see below). Of course, this requirement adds to the computational cost of the quantum calculations.

Arguably, the most effective QM/MM approach for the treatment of enzymatic reactions is the empirical valence bond (EVB) method[18]. The basic idea in the EVB treatment is to obtain the relevant ground state potential in terms of the mixing between the relevant resonance structures (or, more precisely, diabatic states), where each resonance structure represents a specific bonding and charge configuration that has a well-defined interaction with the surrounding environment (protein and/or water).

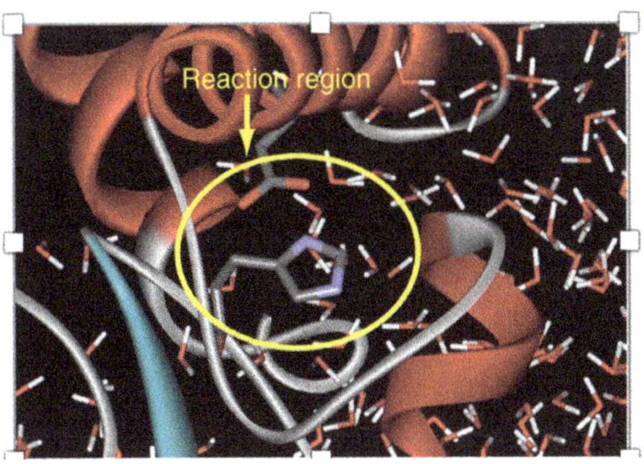

Fig. I.5 The QM/MM model for the enzyme trypsin. The inner circle contains the part that is treated quantum mechanically, while the rest of the system is described by a classical force field. However, the electrostatic potential from the external part is included in the Hamiltonian of the quantum region. This approach allows one to study chemical processes in very large molecular systems. Reproduced with permission[652].

The ground state surface is obtained by solving the secular equation:

$$\mathbf{H}\mathbf{C}_g = E_g\mathbf{C}_g \tag{I.17}$$

where **H** is the EVB Hamiltonian of the system and $\mathbf{C}_g$ is the ground state eigenvector. The matrix elements of **H** are described by analytical functions that can be calibrated by using experimental and theoretical data[18,20] and/or by fitting the calculated ground state surfaces and charge distribution to *ab initio* results[20–24].

The solvent contributions to the Hamiltonian are then considered by evaluating the 'classical' interaction between the solvent and the solute atoms in the different resonance structures (we do not give more details here since the method was described in many works, e.g., Refs. 18, 25).

The EVB free energy surface is evaluated by using a free energy perturbation umbrella sampling procedure, which drives the system from the reactant to the product state and then determines the probability of finding the system at different points along the generalized reaction coordinate[18,25].

The EVB offers what is perhaps the best way of fitting analytical surfaces to *ab initio* surfaces (see, for example, Refs. 21, 26, 27), rather than producing them from first principles. That is, the asymptotic limit of each resonance structure (where the corresponding fragments are at infinite separation) corresponds to a very well-defined molecular fragment that can be reliably described by analytical potential functions (as is now clear from the success of classical force field descriptions of molecules). With the correct physics at the asymptotes, it is usually not difficult to obtain a reasonably accurate description of intermediate regions. This should, of course, be done with the proper off-diagonal mixing terms and a systematic fitting procedure (see below). Perhaps, more importantly, since the EVB incorporates the effect of the solvent in a particularly consistent way, it allows one to "extrapolate" what is known in one environment (e.g., the surface in water) to what we are interested in (e.g., the surface in a protein).

Although the EVB has been found to be extremely effective, it involves the *ad hoc* assumption that the off-diagonal element is similar in the gas phase and in solution, or more importantly in the enzyme and in solution. Now, this assumption is clearly reasonable since it reproduces the well-known linear free energy relationship observed by physical organic chemists for many years[28]. However, a more rigorous validation has clearly been missing. Here, we made a significant advance[29] by using the constrained density functional theory (CDFT) approach. That is, we first defined two diabatic surfaces: one for the reactant and the other for the product (in each surface, we froze the densities according to their values in the corresponding states). Next, we evaluated the free energy functional associated with each state as well as the free energy surface for the ground state DFT surface, which did not involve any constraint density, using Eq. (I.17). The effective H_{12} was then obtained from the two-by-two secular equation, $H_{12}(x) = ((\varepsilon_1 - E_g)(\varepsilon_2 - E_g))^{1/2}$. Repeating this procedure for the reaction in vacuum and solution gave the $H_{12}(x)$ for each case. Now, as is shown in Ref. 30, H_{12} appears to be very similar in both cases, thus giving very strong support to the main assessment of the EVB approach. Further support for these conclusions comes from related applications of CDFT and from the frozen DFT approach. These methods represent significant enhancements over standard QM/MM treatments. They treat the entire system at the *ab initio* DFT level while keeping the densities of ligands and solvent molecules fixed (or constrained) during the self-consistent field treatment of the metal center and reacting fragments. This approach was used in studies of proton transfer in a system that represented the catalytic center of carbonic anhydrase, and the results obtained were quite encouraging[31]. Subsequently we used the CDFT approach in studies of the redox potential of blue copper proteins[32] and in studies of the auto-dissociation of water in water[33], again obtaining encouraging results. The CDFT approach and closely related approaches are also starting to be used by other research groups[34–38]. More recently, we used the CDFT approach in a special adiabatic representation as a tool for exploring the solvent dependence of the EVB off-diagonal element[30]. More importantly, the diabatic implementation of the CDFT approach offers an exciting approach for *ab initio* calculations of free energy surfaces for reactions in solutions and proteins. In this case, we can explore chemical reactions by using the CDFT diabatic surfaces as mapping potentials, which allows us to evaluate the *ab initio* adiabatic free energy surfaces for reactions in solution and protein environments[18].

I.7 Free Energy Perturbation Calculations

Evaluation of free energies by statistical mechanical approaches is extremely time-consuming, due to the need to sample all the relevant configurational space. Fortunately, it is possible in some cases to obtain meaningful results using perturbation approaches. Such approaches exploit the fact that many important properties depend primarily on local changes within the macromolecule, allowing the effects of the overall macromolecular potential to largely cancel out. Such calculations are usually done by the so-called free-energy perturbation method. This method evaluates the free energy associated with the change of the potential surface from U_1 to U_2 by gradually changing the potential surface using the relationship

$$U_m(\lambda_m) = U_1(1 - \lambda_m) + U_2\lambda_m \tag{I.18}$$

The free-energy increment $\delta G(\lambda_m \rightarrow \lambda_{m'})$ associated with the change of U_m to $U_{m'}$ can be obtained by[20]

$$\exp\{-\delta G(\lambda_m \rightarrow \lambda_{m'})\beta\} = \langle \exp\{-(U_{m'} - U_m)\}\beta\rangle_m \tag{I.19}$$

where $\langle\,\rangle_m$ indicates that the given average is evaluated by propagating trajectories over U_m. The overall free energy change is then obtained by incrementally changing λ_m in equal steps and summing the corresponding δG values:

$$\Delta G(U_1 \rightarrow U_2) = \sum_{m=0}^{n-1} \delta G(\lambda_m \rightarrow \lambda_{m-1}) \tag{I.20}$$

The free energy perturbation approach has been used extensively in studies of free energies of biological systems (e.g., Ref. 39). It must be emphasized here that the convergence of free energy perturbation approaches is quite slow and that obtaining meaningful results requires proper treatment of long-range effects.

I.8 Reaction Rates

It is now generally agreed that the rate constant for chemical reactions in the condensed phase can be written as (see references in Ref. 40):

$$k = \kappa k_{\mathrm{TST}} \tag{I.21}$$

where k_{TST} is the transition-state theory rate constant and κ is the "transmission factor", which accounts for the correction of the non-recrossing. In a multi-dimensional system, the TST rate constant can be expressed as (e.g., Ref. 18)

$$k_{\mathrm{TST}} = \frac{1}{2}\langle|\dot{x}|\rangle_{\mathrm{TS}} \exp\{-\Delta g^{\ddagger}/k_B T\} / \int_{-\infty}^{x^{\ddagger}} \exp\{-\Delta g/k_B T\}\,dx \qquad \text{(I.22)}$$

where $\Delta g(x)$ is the free energy functional that describes the probability of being at different points along the reaction coordinate[18,20], $x^{\ddagger}$ is the value of x at the TS, $\dot{x}$ is the velocity, and $\langle\rangle_{\mathrm{TS}}$ designates an average over TS configurations. Here, $\Delta g^{\ddagger}$ is the difference between $\Delta g(x)$ at the TS and the minimum of Δg. The transmission factor k can be determined by running trajectories at the TS region and monitoring their downhill evolution over time. The time reversal of such trajectories (e.g., Ref. 18) produces the rare reactive trajectories and can be used to evaluate k. This is usually done by a counting approach (e.g., Ref. 18; see also Ref. 20). However, we will use an alternative approach below.

The factor $\exp(-\Delta G^{\ddagger}/k_B T)$ in the TST expression for the rate constant is simply a probability factor that could be evaluated by a non-dynamical approach such as Monte Carlo simulations. Thus, if we can estimate the rate constant correctly by Monte Carlo approaches, there are no dynamical effects. Furthermore, it is generally accepted that any dynamical effects on the rate must lie in the transmission coefficient.

I.9 Coarse-Grained Modeling

In many cases we would like to deal with a very large biological system, where all-atom simulations might require enormous computer time for convergence. In such cases it is important to find a simplification that

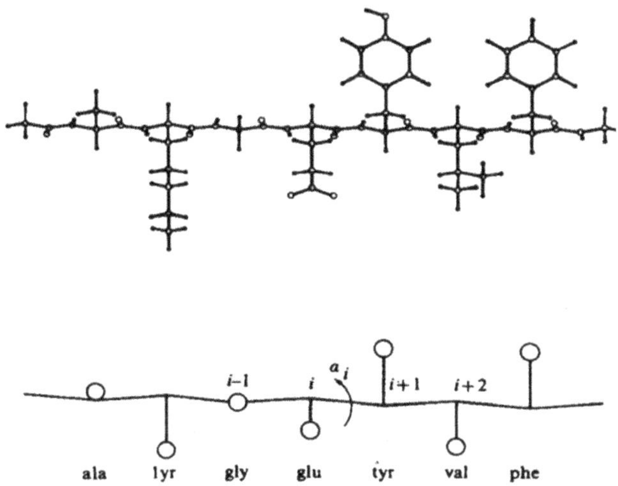

Fig. I.6 The simplified folding model. Reproduced with permission[41].

still retains the main physics of the system. A very useful approximation was proposed in 1975[41] by a simplified folding model (Fig. I.6) that replaced the protein side chains by spheres. These coarse-grained models have been very successful in showing that proteins — with their many degrees of freedom — can reach a folded structure once the motions that are not relevant to folding are averaged.

In subsequent years (e.g., see Chapter **XVII**), we extended the capabilities of the coarse-grained model by making its energy function more reliable through the addition of carefully calibrated electrostatic and hydrophobic terms (described schematically in Fig. I.7). This allowed the model to be applied to studies of protein stability and the behavior of proteins in membranes.

Various cases in which the coarse-grained model proved successful are described in subsequent chapters.

$$\Delta G_{self} = \sum_i \left(U^{np}_{self}\left(N^{np}_i\right) + U^{p}_{self}\left(N^{p}_i\right) + U^{mem}_{self}\left(N^{mem}_i\right) \right)$$

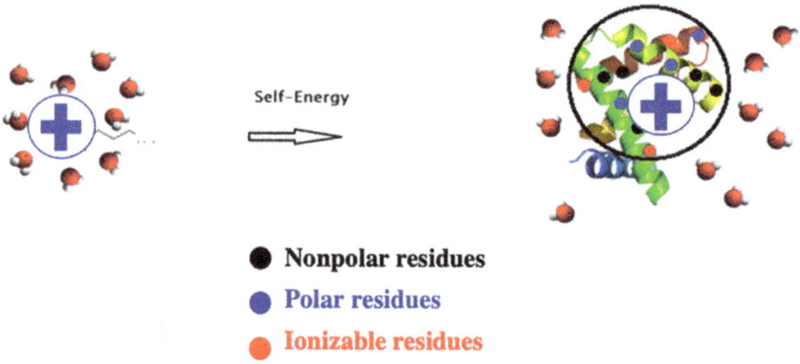

Fig. I.7 The electrostatic-based coarse-grained model. The figure emphasizes the term that evaluates the self-energy in an implicit way.

I.10 Langevin Dynamics/Renormalization

Despite the continual increase in computational power — which has extended the upper limit of simulation time scales to several hundred nanoseconds for condensed-phase systems — many interesting processes remain too slow to be meaningfully simulated using molecular dynamics approaches. However, some of these slow processes can be addressed using Brownian dynamics simulations, which represent the motion of the particles as a stochastic (random) process governed by a modified Langevin equation:

$$m_i \dot{\mathbf{r}}_i = -\gamma m_i \dot{\mathbf{r}}_i + \mathbf{f}_i + (t) + \mathbf{R}_i(t). \tag{I.23}$$

where $\mathbf{R}_i(t)$ is the random force on the ith particle, y is the friction coefficient representing the dissipation of the random forces, and $\mathbf{f}_i(t)$ is the part of the interparticle forces exerted on particle i that are not represented by the random-force and friction terms. The average magnitude of the random forces and the corresponding friction acting on them are connected through the fluctuation-dissipation theorem, with the relationship

$$\langle \mathbf{R}^2 \rangle = n2mkT\gamma \tag{I.24}$$

where n is the dimensionality of the system. At very long times, the behavior described by the Langevin equation reduces to Einstein's diffusion law. The resulting equation of motion is:

$$\mathbf{r}_i\left(t+\delta t\right)=\mathbf{r}_i(t)+\frac{\mathbf{f}_i(t)}{m_i\gamma}\delta t+\rho_i(\delta t) \tag{I.25}$$

This equation no longer involves velocities and represents a coarse-grained view of the time evolution of the system. Here, ρ_i represents the random displacement during a time step of duration δ_i, and ρ is related to the friction coefficient by:

$$\left\langle \rho^2 \right\rangle=\frac{n2kT}{m\gamma}\delta t \tag{I.26}$$

In addition to simulating slow processes[42], Brownian dynamics simulation strategies can also be used for studying general aspects of dynamics[43] or testing theories that often contain assumptions similar to those used in obtaining the Brownian equations of motion[43].

A very useful advance in long time simulation is provided by the **renormalization method**[44], which connects the dynamics of large explicit systems with the dynamics of the corresponding reduced-dimensional systems. In many cases, we aim to study processes that occur over very long time scales — processes that would otherwise require prohibitively long simulations of the full explicit system. In such situations, we can capture the molecular dynamics of the explicit system by running Langevin dynamics on a reduced-dimensional system. The main challenge is obtaining the relevant friction constant. Here, it would not be useful to use standard approaches where the friction is obtained from the velocity autocorrelation since the system can involve many small barriers that may not be captured by simplified formulation. Our resolution to this problem is to force the explicit and simulated systems to have similar behaviors by applying a series of strong forces on the full system and the reduced system and to force both systems to have the same response to the forces by adjusting the friction and the barrier in the Langevin dynamics treatment. Once the appropriate friction is determined, the external forces are removed, and a long Langevin dynamics simulation is run to explore the system's dynamics.

I.11 Time-Dependent Monte Carlo

A very useful approach for long-timescale simulations is provided by **time-dependent Monte Carlo** methods. That is, in principle, the probability of moving over the conformational landscape can be determined by Monte Carlo approaches. However, the associated time scale of such motions is not immediately obvious. In many cases, though, the time dependence can be estimated using alternative methods. Perhaps the best example is presented by proton transfer reactions[45]. In such cases, we can simulate the time it takes for a proton to move between a typical donor and acceptor, which is approximately 1.7 picoseconds in the absence of a barrier. When the transfer involves a free energy barrier, we can use

$$\tau^{-1}(i \rightarrow j) = 0.6 \cdot S \cdot N_{i \rightarrow j} \tag{I.27}$$

where $N_{i \to j}$ is the number of Monte Carlo steps required to move from i to j, S is a scaling factor (discussed below), and 0.6 ps^{-1} is the inverse of the average time it takes a productive trajectory to reach the TS. The derivation of this relationship exploits the fact that, in rate processes, the reacting atoms fluctuate randomly for a long time until a productive trajectory occurs. The overall transfer time can therefore be expressed as the product of (i) the probability of generating a productive trajectory and (ii) the time it takes for that trajectory to cross the TS. Thus, we use:

$$S = \exp\left\{\frac{\Delta g^{\ddagger}_{i \to j} - \Delta G_{i \to j}}{k_B T}\right\} \tag{I.28}$$

Chapter II

Basis of Electrostatics

Static electricity was first described by Thales of Miletus around 600 BC, and the dependence of electrostatic force on distance was formally established in 1785 by Charles-Augustin de Coulomb. Finally, the general laws of electrodynamics were formulated by James Clerk Maxwell (Fig. II.1) in 1861.

The major advances in electrostatic theory were developed before the discovery of quantum mechanics and before the molecular nature of matter was fully understood. As a result, although this chapter addresses a central theme of the book — namely, the treatment of electrostatics — it takes a different approach than most standard electrostatics textbooks. Rather than focusing on a purely macroscopic framework, this chapter emphasizes the connection between the molecular (microscopic) picture and the corresponding macroscopic descriptions. Thus, the emphasis here is placed on developing a clear microscopic understanding, as opposed to relying on macroscopic models in which the dielectric constant serves as a poorly defined parameter — one that is clearly inadequate for describing energetics within the interior of proteins.

Fig. II.1 James Clerk Maxwell.

II.1 Basic Formulation

The starting point for the formulation of electrostatic theory was Coulomb's law. This foundational observation established that the reversible work (or free energy) required to bring two charges (Q_i and Q_j) together **in vacuum** is given by:

$$\Delta G = 332\ Q_i Q_j / r_{ij} \tag{II.1}$$

where the free energy is expressed in kcal/mol, the distance r_{ij} is in Å, and the charges Q are in atomic units (au). Manipulations of this equation led to the general macroscopic form of the electrostatic field equation[46]:

$$\nabla \mathbf{E}(\mathbf{r}) = -4\pi\rho \tag{II.2}$$

where $\mathbf{E}$ is the macroscopic electric field and ρ is the charge density $\left(\rho(\mathbf{r}) = \Sigma_i Q_i \left(\delta(\mathbf{r} - \mathbf{r}_i)\right)\right)$. Further manipulation, using the fact that the field **E** can be expressed as a gradient of a scalar potential U, gives:

$$\mathbf{E} = -\nabla U \tag{II.3}$$

which leads to the Poisson equation:

$$\nabla^2 U(\mathbf{r}) = -4\pi\rho(\mathbf{r}) \tag{II.4}$$

Assuming the system can be represented by a dielectric constant ε, this becomes:

$$\nabla\varepsilon(\mathbf{r})\ \nabla U(\mathbf{r}) = -4\pi\rho(\mathbf{r}) \tag{II.5}$$

When ions are immersed in the solution surrounding our system and the ion distribution follows the Boltzmann distribution, we obtain an approximation known as the linearized Poisson–Boltzmann (PB) equation:

$$\nabla\varepsilon(\mathbf{r})\nabla \mathrm{U}(\mathbf{r}) = -4\pi\rho(\mathbf{r}) + \kappa^2 U \tag{II.6}$$

where κ is the Debye-Hückel screening parameter.

The above treatment has one major limitation: it relies on the problematic assumption that the medium can be represented as a structureless continuum. However, for a more fundamental physical understanding, it is essential to develop a more detailed, microscopic description. The starting point for this improved perspective is the relationship between microscopic and macroscopic fields.

II.2 Energies and Field

The microscopic potential energy of a collection of charges is given by

$$V = \frac{332}{2}\sum_{ij} Q_i Q_j / r_{ij} + V_\infty \tag{II.7}$$

where V_∞ is the energy of the system when all charges are infinitely separated. Here, V is given in kcal/mol, r is in angstroms (Å), and the charges Q in atomic units (au). The force exerted by point charges on a unit test charge at position **r** defines the microscopic electric field, which is given (in kcal/mol/Å²) by:

$$\varepsilon(\mathbf{r}_i) = 166 \sum_j Q_j \mathbf{r}_{ij} / r_{ij}^3 \tag{II.8}$$

In formulating the electrostatic energies we should also consider the dipole moments defined by

$$\mu_i = I_i q_i$$

where I_i is the vector between two opposing charges q_i and $-q_i$.

The potential energy of a dipole μ_i in a field $\varepsilon(\mathbf{r}_i)$ is given by

$$U = -\mu_i \varepsilon(\mathbf{r}_i) \tag{II.9}$$

Of particular interest is the case where the electron cloud of each molecule is polarized by the field from surrounding molecules, leading to the formation of an induced dipole:

$$\mu_i = \alpha \varepsilon(\mathbf{r}_i) \tag{II.10}$$

where the proportionality constant α is the molecular polarizability.

In the case of a collection of polarizable particles (e.g., rare gas atoms) that are subjected to a uniform external field ε^0, the induced dipole (i.e., the polarized electron cloud of each particle) responds to the local field $\varepsilon(\mathbf{r}_i)$. This local field is the sum of the external field ε^0 and the field from all other dipoles at $\mathbf{r}_i$:

$$\mu_i = \alpha\left(\varepsilon^0 + \varepsilon_\mu\right) = \alpha\left[\varepsilon^0 - \sum_{j \neq i}\left(I/r_{ij}^3\right)\left(\mu_j - 3\left(r_{ij}\mu_j\right) r_{ij} / r_{ij}^2\right)\right] \tag{II.11}$$

Here, ε_{μ} is the field dipole i due to all other dipoles (excluding its own self-field). This expression can be reformulated in matrix notation as:

$$\boldsymbol{\mu} = (\mathbf{I} - \alpha\mathbf{B})^{-1}\alpha\varepsilon^{0} \quad \text{(II.12)}$$

where **B** is the diagonal matrix of the atomic polarizabilities, and the elements of matrix **B** are given by:

$$B_{ij}^{st} = \left[\left(\delta^{st} - 3\mathbf{r}_{ij}^{s}\mathbf{r}_{ij}^{t}/\mathbf{r}_{ij}^{2}\right)/\mathbf{r}_{ij}^{3}\right]\left(1 - \delta_{ij}\right), \quad \text{(II.13)}$$

Here, s and t run over the x, y, z components of each vector. Instead of solving Eq. (II.12) directly, it is more convenient in many cases to solve Eq. (II.11) iteratively: first calculating the dipoles due to ε^{0}, then computing the local field by including the dipole-dipole interactions (and updating the dipoles accordingly). The new dipoles induced by the new local field are then used for the next iteration. Implementing this approach forms the basis of the so-called "polarizable force field"[10,47].

It is worth noting that classical microscopic treatments of the energetics of induced dipoles in solutions and large molecules only began to emerge in the mid-1970s. A preliminary attempt to study dielectric effects in nonpolar environments was made by Hopfinger[48], who modeled a methyl group placed between two charges. However, this study overlooked a key physical insight: most dielectric effects arise from the molecules surrounding the charges, not from those located directly between them. Thus, the first consistent microscopic treatment of dielectric effects in polar/nonpolar environments was reported in our work[10], in which we simulated the electrostatic environment in lysozyme using a classical polarizable force field. The surrounding solvent was modeled using a grid of Langevin-type dipoles[10]. Similar approaches were used for other proteins (e.g., Ref. 49) and for polarizable grids of dipoles (e.g., Ref. 3). Thus, the use of polarizable force fields dates back to the work of Warshel and Levitt[10], who introduced this approach as a general framework for capturing the effect of electronic polarization and the corresponding dielectric constant in protein modeling. This was done using both iterative and non-iterative approaches. Subsequent early instructive studies include those reported in Refs. 50, 51.

The macroscopic description of electrostatic fields is quite different from the microscopic description considered above. Microscopic evaluation of electrostatic fields involves a straightforward discrete summation of the field from each molecule in the system. On the other hand, the macroscopic approach (which was developed before the elucidation of the atomic nature of matter) treats the medium as a continuum. The field inside this continuum includes contributions from all charges, including the field from an atom on itself. Such an approach might seem irrelevant to those who are interested in simulating detailed molecular interactions where the field from a molecule on itself goes to infinity and does not lead to any useful insight.

Yet, despite the focus of this chapter, nearly all literature on electrostatic theory relies on the macroscopic approach — and this formalism has led to substantial scientific progress. Thus, it is important to understand the relation between the microscopic and macroscopic fields. To clarify this relationship, we consider a very large cube of polarizable material subjected to an external electric field (Fig. II.2). We then remove a small region — denoted as region (I) — from this large system and fix the dipoles within this region at the same magnitudes and orientations they had in the fully polarized system. Next, we attempt to evaluate the electric field at an arbitrary point within Region I. However, in

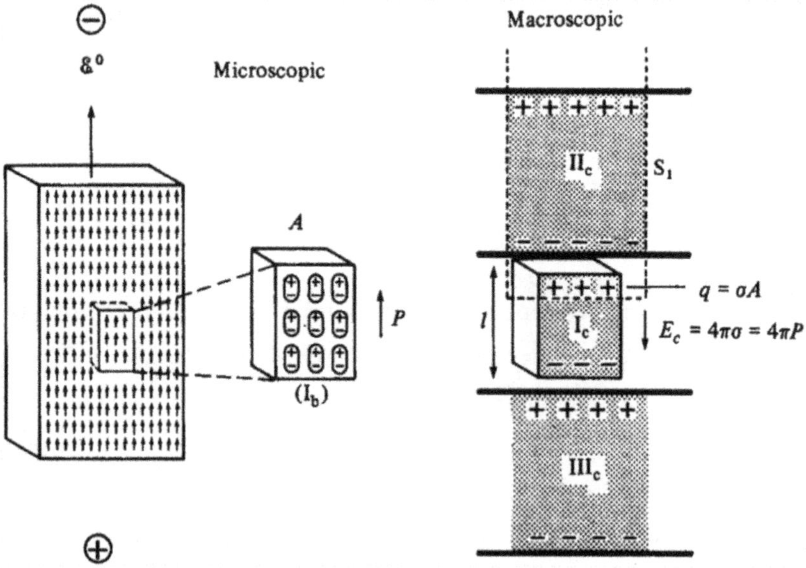

Fig. II.2 The relation between microscopic and macroscopic fields. The figure and the corresponding discussion in the text are used to prove that the macroscopic field E is given by the average of the microscopic field ε (see the derivation of Eq. (II.17)). Reproduced with permission[52].

contrast to the microscopic approach, we now include the contributions from all dipoles — including the dipole at the point of interest itself. This modified field is denoted ε'_μ, as opposed to ε_μ in the microscopic case. That is, we allow the self-field to contribute. The result (illustrated in Fig. II.3) is an electric field that oscillates wildly and diverges to infinity at molecular sites.

Next, let us try to evaluate the field at any arbitrary point inside region (I). Now, however, in contrast to the microscopic treatment, we include in the field the contributions of all dipoles (this field is called ε'_μ rather than ε_μ). While the evaluation of this oscillating field seems quite useless, the average of the field does, in fact, converge to a *constant*

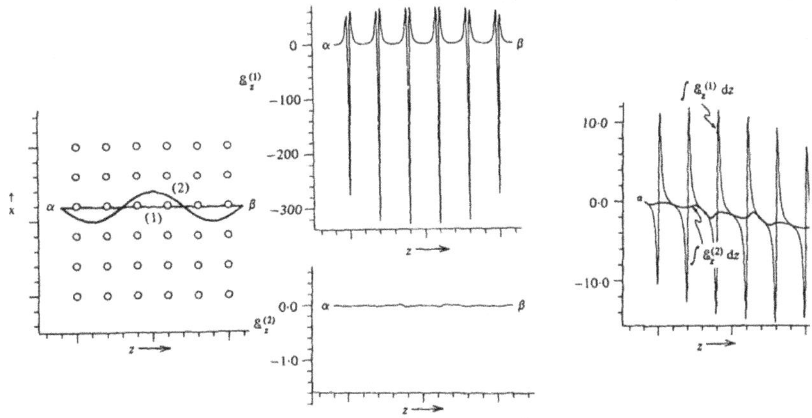

Fig. II.3 The integral of the field across a polarized medium is independent of the pathway. As discussed in the text, although the field across different pathways can be entirely different, the integral of the field between two given points is independent of the pathway. Reproduced with permission[52].

limit. To find this limit, we use the fact (see Fig. II.3) that the path integral of the field

$$\left(\phi_{\alpha\beta} = \int_{\alpha}^{\beta} \varepsilon'_{\mu} \mathrm{ds}\right) \tag{II.14}$$

between any two points α and β is the corresponding electrostatic potential, which is independent of the pathway (even if it passes through molecules). Since this integral is independent of the path, we could try to evaluate it for points outside region (I). The simplest way to do this is to use the fact that the field outside the region is the same as the field of an equivalent cube with charges $+\sigma$A and $-\sigma$A on its top and bottom surfaces, where A is the area of the surface of region (I) and σ is the charge density per unit area (the reason is clear: one compares the net charge distribution of the microscopic and macroscopic representations of region (I) in Fig. II.2). Since the path integral between the points on both sides of the macroscopic equivalent of region (I) (referred to as (lc)) is equal to the corresponding path integral for microscopic region (I), we obtain:

$$\int_b \varepsilon'_\mu \mathrm{ds} = \int_c \varepsilon'_c \mathrm{ds} \tag{II.15}$$

where *b* and *c* are, respectively, the microscopic and macroscopic representations of region (I) and ε'_c is the microscopic field obtained from the charge distribution of region (Ic). Integrating this equation across lines that pass parallel to ε^0 through equally spaced points on the upper surface of region (I) we obtain:

$$\frac{1}{\tau}\int_h \varepsilon'_\mu \mathrm{ds}dA = \frac{1}{\tau}\int_h \varepsilon'_\mu \mathbf{e}d\tau = \left\langle \varepsilon'_\mu \mathbf{e} \right\rangle_\tau = \frac{1}{\tau}\int_\mu \varepsilon'_c \mathbf{e}d\tau = \mathbf{E}_\mu \mathbf{e} \tag{II.16}$$

where **e** is a unit vector in the direction of ε^0, τ is the volume of region (I), and **E** is the *macroscopic* field due to the dipoles of region (I). Since, as will be shown below, the dipoles in regions (II) and (III) do not contribute to the field in region (I), we finally obtain the key relation

$$\mathbf{E} = \left\langle \varepsilon^0 + \varepsilon'_\mu \right\rangle_\tau = \left\langle \varepsilon' \right\rangle_\tau \tag{II.17}$$

That is, the average of the field $\left(\varepsilon^0 + \varepsilon'_\mu\right)$ in a macroscopically small region is the corresponding macroscopic field. The size of the microscopic region needed for a meaningful average is considered in Fig. II.3.

Now we can evaluate the macroscopic field, E, using the powerful Gauss's theorem. This theorem states that the integral of the field over the surface that encloses a given region in space can be evaluated by

$$\oint \mathbf{En}da = 4\pi q \tag{II.18}$$

where **n** is the vector perpendicular to the surface element *da* of the given region, and *q* is the net charge within the enclosed area. Using this theory, it is easy to evaluate the macroscopic field $\mathbf{E}_c$ in region (I) of Fig. II.2 due to all the dipoles of the system. Integrating **E** over the surface S_1 in the figure gives contributions only from the field inside region (I) (as the upper part of S_1 is taken to infinity). Since the contribution from region (II) is zero and the overall charge in S_1 is given by σA (see Fig. II.2), we obtain

$$\oint \mathbf{E}_c \mathbf{n} da = E_c A = 4\pi\sigma A. \tag{II.19}$$

As seen from Fig. II.1, the total dipole moment of region (I) is given by both the sum of the individual dipoles and by the dipole moment obtained by the product of the charge *q* and the length *l*. Thus, the average dipole per unit volume — defined as the polarization vector **P** — has a magnitude given by:

$$\mathbf{P} = \langle \mu \rangle_\tau = ql/\tau = \sigma Al/Al = \sigma \tag{II.20}$$

Equations (II.19) and (II.20) give

$$\mathbf{E}_c = \langle \mathbf{E}_c \rangle_\tau = 4\pi\mathbf{P} \tag{II.21}$$

or just

$$\mathbf{E}_c = 4\pi\mathbf{P} \tag{II.22}$$

The macroscopic field $\mathbf{E}_c$ is the field due to the dipoles in the volume element (I). The total macroscopic field in region (I) does not include additional contributions from the dipoles outside this region. This can be shown by considering the fact that the dipoles from region (II) do not contribute to the charge used in evaluating $\mathbf{E}_c$ by Gauss's theorem. Thus, the total field **E** is the average of the field from the dipoles inside the given volume element and the external field ε.

Because the induced field is directed opposite to the external field (since **P** is parallel to ε^0), we obtain:

$$\mathbf{E} = \mathbf{E}^0 + \mathbf{E}_c = \mathbf{E}^0 - 4\pi\mathbf{P} \tag{II.23}$$

Equation (II.23) was derived for the case considered in Fig. II.2 (where the field from region (II) at region (I) is zero). For more general cases it is customary to replace $\mathbf{E}^0$ by the so-called 'electric displacement vector' defined by:

$$\mathbf{D} = \mathbf{E} + 4\pi\mathbf{P} \tag{II.24}$$

It should be mentioned, however, that this widely used definition is virtually useless for practical studies, in which **D** and **E** are both unknown and **D** should be evaluated in terms of **E**. If **D** is parallel to **E** we can write

$$\mathbf{D} = \epsilon\mathbf{E} \tag{II.25}$$

where ϵ is called a dielectric constant. From Eqs. (II.24) and (II.25) one obtains:

$$4\pi\mathbf{P} = (\epsilon - I)\mathbf{E} \tag{II.26}$$

In general, ϵ in Eq. (II.25) should be a non-constant matrix, and its molecular meaning is an important point in this work.

The interest in the microscopic basis of the dielectric effect led to landmark works[53,54] that explored the relation between the macroscopic dielectric constant and the dipole moments of the molecules of the given system. Despite the enormous impact of these works, they did not provide practical ways for quantitative evaluation of the electrostatic energy of charges in solutions. This problem probably cannot be solved without the use of high-speed computers. This point can be seen using the results of Fig. II.4. This figure gives the macroscopic field and dielectric constant, for a system of a charge in a polarizable fluid, as a function of the distance from the charge. Using the calculated **E** and **P** as well as Eqs. (II.11) and (II.24), one obtains the corresponding $\epsilon(r)$ (see Fig. II.4). As seen from the figure, the function $\epsilon(r)$ starts to approach the value of the bulk dielectric constant only when the volume element used for averaging is sufficiently large. More systematic analysis of dielectric constants will be provided in Section IV.3.

At this point, it may be interesting to share a brief personal perspective. Since my undergraduate studies, I have been unsatisfied with the typical textbook explanation for the appearance of the factor of one-half in the energy expression describing the interaction between an induced dipole and an external field. The idea that matter can be represented by induced dipoles goes back to early electrostatics literature. However, the proper microscopic description of polarization — and the replacement of electronic polarization with classical polarizable induced dipoles — is a more recent development. Most textbooks present the energetics of polarizable matter using a macroscopic treatment, yet the connection to the microscopic description is often unclear. For example, according to well-established macroscopic theory (e.g., Refs. 55, 56), one can express the energy of a polarizable volume element by:

$$W = -\frac{1}{2}\mathbf{P}\mathbf{E}_0 = -\frac{1}{2}\alpha\mathbf{E}_0^{\,2} \tag{II.27}$$

where **P** is the induced polarization, $\mathbf{E}_0$ is the macroscopic field, and α is the corresponding polarizability.

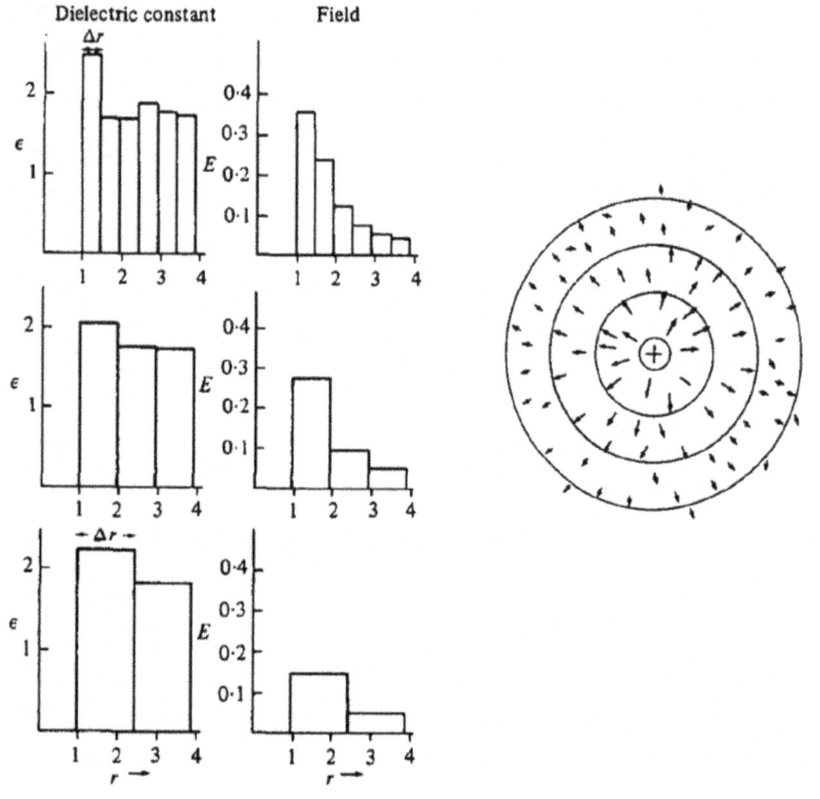

Fig. II.4 Averaging microscopic fields and dipoles gives the macroscopic field and dielectric constant when the volume element used is sufficiently large. The figure evaluates the macroscopic field and dielectric constant for a polarizable fluid around an ion. The calculations demonstrate how the macroscopic field $\left(E=\left\langle \varepsilon^{0}+\varepsilon'_{\mu}\right\rangle\right)$ and macroscopic dielectric constant $\epsilon(r) = (\langle E\rangle + 4\pi\langle P\rangle)/\langle E\rangle$ depend on the volume element used for the averaging of the corresponding microscopic field. The results shown were obtained by evaluating the microscopic induced dipoles for a polarizable fluid around a charge, and then calculating the field, E(r), and polarization, P(r), by averaging the corresponding microscopic field and microscopic dipoles for the spherical region between r and r + Δr. As shown in the figure, the dielectric constant approaches its bulk value once the volume elements are sufficiently large. The calculations were done with a grid spacing of 1 Å and polarizabilities of $\alpha = 0.05\text{Å}^3$ which corresponds (according to the Clausius-Mossoti equation) to a bulk dielectric constant of $\epsilon = 2$. Reproduced with permission[52].

In molecular systems, it is well established that the interaction energy between a charge and the induced dipole of a single atom is given by: $W = -\mu\xi_0 = -\alpha\xi_0^2$ (where ξ is the microscopic field on the atom). Thus, the origin of the (1/2) factor in Eq. (II.27) was not obvious. When I raised this question — asking professors in physics or electrical engineering how the factor arises — the usual response involved invoking the well-known macroscopic integral $\int QdQ = Q_0^2/2$ or, more generally, arguments about the linear response nature of matter. However, this type of explanation never fully satisfied me, and I insisted on a more molecular-level understanding. My curiosity deepened when I came across Boettcher's book[56], where he also questioned the validity of this 1/2 relationship. Boettcher begins by stating that in a system of real, discrete charges, the 1/2 factor does not appear (as seen in his Eq. (3.24)). He then transitions into a continuum model and introduces the integral $\int QdQ$ to recover the 1/2. Finally, in his Eq. (3.30), he attempts to reconcile the continuum and discrete descriptions, concluding that the charging energy must be adjusted by including the 1/2 factor. This is followed by a rather convoluted discussion (below his Eq. (3.30)) involving the concept of infinitely large energy. Despite the above effort, I could not adopt this explanation as the foundation for my own theoretical work. Eventually, I clarified the origin of the 1/2 factor by considering the polarizable medium as a collection of explicit atoms with polarizable electron clouds. This approach made it possible to rigorously describe the microscopic relationship for a system of interacting charges and induced dipoles[10]. In this framework, the total energy of interaction can be expressed as:

$$W = -\sum_{i,j} Q_i\left(\mu_j \cdot \mathbf{r}_{ij}\right)/r_{ij}^3 + \sum_{j>j'} \mu_j\left[\nabla\left(\mu_{j'}\mathbf{r}_{j'j}\right)/r_{j'j}^3\right] + \frac{1}{2}\sum_j \alpha_j\left|\xi_j\right|^2 \tag{II.28}$$

Here, the first term comes from the interaction between the charges Q_i and the dipoles μ_j, the second term accounts for the interaction of dipoles j and j', and the third term represents the energy that must be spent in distorting the electron cloud of the atom to create the induced dipoles. The corresponding energy cost was verified by using a model that views the electron as being attached to the nuclear core by a spring and by actual quantum mechanical calculations which consider an atom in an external field. Ultimately, Eq. (II.28) can be written as[10]

$$W = -\frac{1}{2}\sum_{i,j} Q_i \left(\mu_j \cdot \mathbf{r}_{ij}\right) / r_{ij}^3, \tag{II.29}$$

Or equivalently (see also Ref. 52):

$$W = -\frac{1}{2}\sum_{j} \mu_j \cdot \xi_j^0, \tag{II.30}$$

where ξ_j^0 is the field acting on the jth dipole from the charges in the system. This field does not include the field from the other dipoles; that leads, however, to the actual value of μ_j. The above derivation has not appeared, to the best of my knowledge, in the early macroscopic literature.

In general, the clearest way to understand the origin of the 1/2 factor in molecular systems is to recognize that there is no such factor in the direct interaction between a dipole and a charge. Instead, the factor of 1/2 emerged when we evaluate the energy required to create the dipoles — not using a continuum approximation, but through a quantum mechanical treatment or a spring model for the electron. This approach reveals that the energy needed to induce the dipoles is equal to half the interaction energy between the resulting dipole and the field. Ultimately, the issue of the half factor serves as an example of my difficulties in trusting macroscopic results in my attempt to understand enzymes and related systems. This led me to conclude that a more reliable strategy was to adopt an explicit, yet simplified, representation of the surrounding environment.

Chapter III

Electrostatic Models

Dealing with electrostatics in complex systems such as proteins cannot be accomplished through analytical formulations alone and instead requires the use of numerical computer models. These models must be both practical — in terms of computational efficiency — and sufficiently accurate to reproduce experimental observations and to provide insight into the origins of biological function. In this chapter, we will consider the development and scope of such models.

Progress in studying electrostatic effects in proteins has involved a wide range of modeling approaches, which can be roughly categorized into microscopic all-atom models, simplified dipolar models, and macroscopic continuum models. Each of these frameworks has its own scope, strengths, and limitations. This includes, of course, the difficulty of analyzing continuum results without the use of microscopic models. Thus, it is important to understand the relationship between various approaches, spanning a range from continuum dielectrics to all-atom classical representations and quantum mechanical models. In the sections that follow, we will examine the key classes of electrostatic models and discuss their underlying assumptions, capabilities, and current status. A personal perspective on the development of some of these models will be offered at the end of the chapter.

III.1 All-Atom Models, Boundary Conditions, Long-Range Treatments, Polarizability Effects, and Convergence

All-atom simulations describe molecular systems as collections of particles that interact via a quantum mechanical potential surface or the corresponding classical force field. While these approaches can be quite rigorous, sufficient convergence may not be achieved (even with nanosecond/microsecond simulations), particularly in the case of macromolecules. In this respect, one of the most significant issues in atomistic simulations is the proper representation of an infinite system, which is critical for the proper treatment of long-range electrostatic effects. A seemingly appealing solution to this problem is the use of periodic boundary conditions.

However, although this approach is appropriate for treating nonpolar systems and for studies of non-electrostatic problems (or in studying molecular crystals[57]), periodic boundary conditions cannot produce truly infinite non-crystalline systems without artifacts when long-range electrostatic interactions are significant (the problems are described schematically in Fig. III.1). In fact, it has been long recognized that the Ewald method, commonly used with periodic boundary conditions, yield divergent solvation energies for the same solute depending on the size of the simulated system[58]. Correction formulas do exist for specific charge distributions[59], but they do not provide a general solution (see discussion

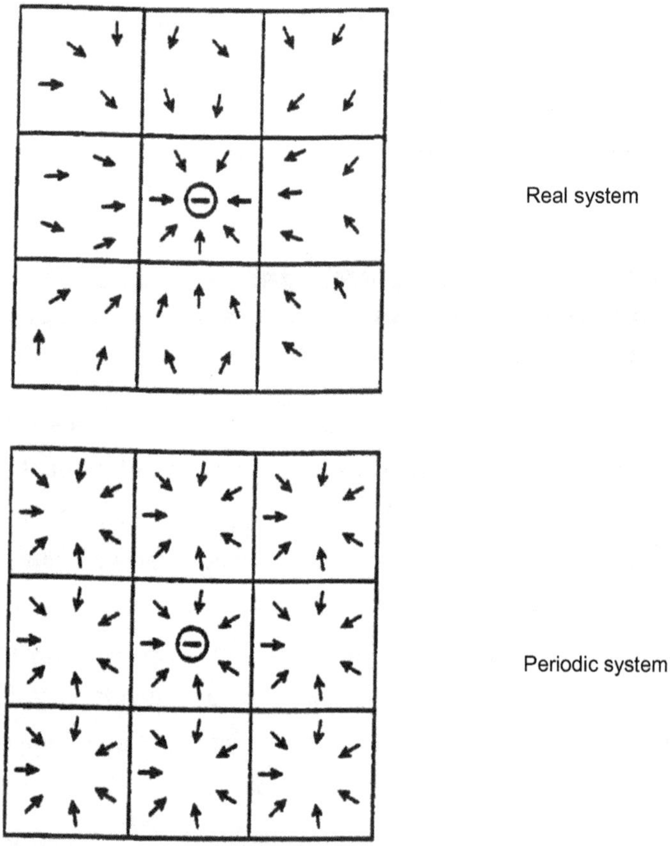

Fig. III.1 Problems associated with the use of periodic boundary conditions for studies of solvation of ions. Reproduced with permission[655].

in Refs. 60 and 61). While the use of very large unit cells can reduce these artifacts, such an approach dramatically increases computational cost. Nevertheless, Ewald-based methods have become standard components of many modern simulation packages, often giving the impression that long-range electrostatic issues are minor or resolved. In practice, however (especially under limited computational resources), it is important to explore more effective models with better physical bases. Here (as demonstrated in several studies[60,62,63]), it is useful to consider advanced simulations that use spherical boundary conditions and utilize the local reaction field (LRF) long-range treatment[64]. Spherical boundary conditions offer an obvious alternative that is more physically consistent than the Ewald methods. There is no spurious periodicity; a spherical region of atomic representation is surrounded by a simplified region that represents the rest of the system. Because the simplified outer region is far from the focal point of the simulation, its effect can be made to be indistinguishable from that of an atomistic representation. The issues involved in constructing proper boundaries for spherical models are logical and obvious. The most critical requirement is ensuring that the system maintains proper polarization at and around the boundaries[3].

This requirement, which is far more important than the physically trivial constraint of keeping proper heat flow by stochastic boundaries[65], has been achieved by the surface constraint all-atom solvent (SCAAS) model[65] (see Fig. III.2) and by the earlier surface constrained soft sphere dipole (SCSSD) model[3].

The problem is not the well-known bulk contribution from the surrounding of the simulation sphere[66], which has been implemented in some simulation studies (e.g., Refs. 67, 68), but rather the polarization at the microscopic boundaries of the simulation sphere — an effect that has been recognized and implemented in the SCAAS model and in the earlier SCSSD model.

It is important to consider the effect of induced dipoles in microscopic simulations of electrostatic energies in condensed phases. This can be achieved using polarizable force fields. This approach dates back to Ref. 69, which first introduced it as a general framework for capturing

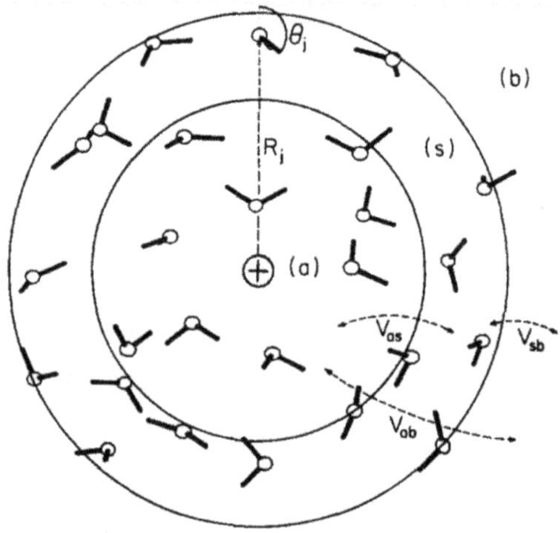

Fig. III.2 An illustration of the three regions of the SCAAS model. Region (a) contains the solute and its first few solvation shells. Region (s) is the surface region. Molecules in this region are subjected to the constraint forces described in the text. Region (b) is the bulk region, which is represented by a continuum model. Reproduced with permission[65].

electronic polarization effects in protein modeling. This was done in both iterative and non-iterative approaches. Recognition of the importance of the effects of indued dipoles in electrostatic modeling developed slowly, but is now widely appreciated[70,71] (for a review, see Ref. 47).

With reasonable boundary conditions, proper long-range treatment, and induced dipoles, one can use all-atom models to evaluate electrostatic free energies. Here, the most obvious approach is the free energy perturbation (FEP) method, in which the solute charge is gradually changed from zero to its actual value $Q = Q_0$, using a mapping potential of the form:

$$V_m = V(Q = 0)(1 - \lambda_m) + V(Q_0) \tag{III.1}$$

This is followed by evaluating the corresponding free energy change using Eq. (I.17).

A useful alternative approach is the linear response approximation (LRA), in which the free energy is given by:

$$\Delta G(Q=0\rightarrow Q=Q_0)=\frac{1}{2}\Big[\big\langle V(Q_0)-V(Q=0)\big\rangle_{V(Q_0)}+\big\langle V(Q_0)-V(Q=0)\big\rangle_{V(Q=0)}\Big] \quad \text{(III.2)}$$

Although both the FEP and LRA approaches are very appealing (e.g., Refs. 72–76), they suffer from significant convergence problems.

Another seemingly rigorous approach for studying electrostatic energies in proteins — particularly for evaluating the free energy profile of ions in ion channels — is the calculation of the so-called potential of mean force (PMF). The PMF reflects the free energy of moving a given charged group from the bulk solvent to a specific protein site. It is typically evaluated using umbrella sampling or related methods[77,78].

In this approach, one can use a mapping potential of the form,

$$\begin{aligned}\varepsilon_m &= (1-\lambda_m)\varepsilon_1+\lambda_m\varepsilon_2\\ \varepsilon_1(z) &= E_g(z)+E_{1,cnstr}(z)=E_g(z)+K\big(z-z_0^{(1)}\big)^2\\ \varepsilon_2(z) &= E_g(z)+E_{2,cnstr}(z)=E_g(z)+K\big(z-z_0^{(2)}\big)^2\end{aligned} \quad \text{(III.3)}$$

where $z_0^{(1)}$ and $z_0^{(2)}$ are the two neighboring points on the z-axis, and $E_g(z)$ is the total potential of the system (i.e., the solvated Na^+ in an ion channel) without the constraints. We thus use the following FEP/umbrella sampling expression[78]:

$$\Delta g(z)=-\frac{1}{\beta}\ln\Big[\exp\{-\Delta G(\lambda_m)\beta\}\big\langle\delta(z'-z)\exp\{E_{m,cons}(z)\beta\}\big\rangle_{\varepsilon_m}\Big] \quad \text{(III.4)}$$

It has been popular to replace the use of the ΔG term with the implementation of the so-called weighted histograms analysis method, which is intended to provide optimal overlap between simulation results obtained at different values of z'. However, it appears that this method does not significantly improve the quality of the results[78].

We highlight this point because the discussion of the seemingly rigorous formulation of PMF methods and related approaches (e.g., Ref. 79) may have drawn attention away from the core issue: the convergence of PMF calculations and their accuracy in estimating the work required to move ions from bulk solvent into the interior of a protein. Furthermore, since PMF approaches do not provide information on absolute solvation free energies, they do not provide a clear indication about possible major problems with the simulation method used with a given treatment of long-range effects.

Instructively, Ref. 78 compared the performance of FEP-based PMF (obtained by charging an ion at different points along the reaction coordinate) to a non-standard PMF obtained by pulling the ion using Eqs. (III.3) and (III.4). The comparison was done for a well-defined benchmark composed of a small part from the gramicidin A channel with different degrees of solvation. The result confirmed the assumption that the regular PMF approach may involve major hysteresis effects.

Nevertheless, it should be mentioned that the major increase in computer power has led to increasingly more reliable microscopic studies of ion flow through ion channels[80,81].

III.2 Dipolar Models Such as the LD and PDLD Models are Simplified Yet Microscopic Models

The most natural simplification of all-atom polar models involves the representation of molecules or atom groups as dipoles. This is in fact the picture that was used as a conceptual starting point in the development of electrostatic theories[82]. Furthermore, dipolar models have a long history in the theory of polar solvents. Simulations with sticky dipoles[83] and formal studies[84] addressing the origin of the dipolar representation have been recently reported. As much as solvation models are concerned, the first advance that used a dipolar model has been the Langevin dipole (LD) model[10], where the water molecule around a solute were represented by a grid of dipoles that are polarized by the Langevin equation, considering the field from the solute and the other dipoles. While the LD model was introduced based on logical but somewhat *ad hoc* considerations, its validity was analyzed by exploring the relationship between this model and the much more rigorous Brownian dipole lattice (BDL; see Fig. III.3), as well as the relationship between these models and continuum models[85]. This study indicated that continuum solvation models are the infinite dipole density limit of a more general dipolar representation, and that their linearity is a natural consequence of being at that limit (although the discreteness of dipole lattice models is not equal to that of continuum models[86]). It was also found that the solvation behavior of the LD model is identical to that of the more rigorous BDL model.

This is significant since a version of the LD model, the protein dipoles Langevin dipoles (PDLD) model, has been used extensively to treat

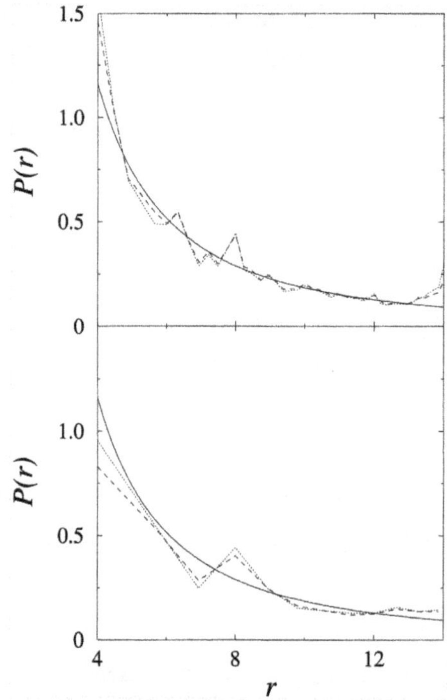

Fig. III.3 Comparison of polarization profiles $P(r)$ in BDL and LD, for $\epsilon \approx 6$, when solvating a unit charge in a 4 Å cavity (formed by deleting dipoles or setting $\epsilon = 1$ at $r < 4$ Å) concentric with the solvent sphere with a radius of 14 Å. $P(r)$ is in 10^{-3} au, and r is in angstroms. The solid curve is the analytical continuum dielectric prediction. Dotted and dashed curves represent $P(r)$ in BDL and LD, respectively. (a) Fine lattice ($a = 2$ Å). (b) Coarse lattice ($a = 4$ Å). Reproduced with permission[85].

electrostatics in macromolecules[87]. Apparently, the PDLD model has been the first to properly treat macromolecular electrostatics.

The PDLD model has been criticized by some workers, who otherwise have little difficulty in accepting macroscopic models as realistic description of macromolecules. This is unfortunate in view of the obvious conclusions of repeated rigorous studies of dipolar lattices[86] and related models (e.g., see Ref. 88). An instructive example is the problematic suggestion[89] that the PDLD model did not evaluate self-energies, and the assertion that this model is, in fact, macroscopic in nature and that it is physically equivalent to the use of a finite-difference grid within discretized continuum (DC) methods. These assertions involve a misunderstanding of the physics of dipolar solvents. It should be clear by now that a grid of dipoles with finite spacing or the equivalent system of polar molecules on a lattice is not equivalent to the numerical grid used in evaluating the electrostatic potential in DC approaches (see the analysis of footnote 26 of Ref. 73). Another example that supported the incorrect argument that the LD and DC models are equivalent was advanced by Ref. 90. This work stated: "The LD model is not only analogous to the continuum model (as stated many times in the literature), but identical to the continuum model in its discrete approximation," continuing "to our knowledge this rigorous equivalence was first pointed out in Ref. 88". Unfortunately, this reflects a misunderstanding of the conclusions of Ref. 88, and also a problematic overlooking of the rigorous findings of Ref. 86, where it was found that the LD and DC models have different structural features even at the limit of infinitely small spacings.

It should be realized by now that the original LD model, with its finite spacing, has always been a simplified microscopic model. Insisting on the clear physical features of the model (i.e., the clear identification of each grid point as a discrete particle) is the reason that the PDLD model has avoided the traps that plagued continuum studies of proteins. For example, looking at the microscopic interactions led immediately to the explicit considerations of the protein permanent dipoles rather than to the problematic attempts of representing them by a low dielectric constant. At any rate, saying that the LD model (or other dipolar model) is equal to a continuum model is like saying that an all-atom model is equal to a continuum model. Finally, if one still likes to consider the suggestion that the LD model is a continuum model[89,90], it is useful to point out that macroscopic models involve scaling by a dielectric constant but no such scaling is used in the PDLD model.

III.3 Consistent Link between Microscopic and Semi-Microscopic Treatment; the PDLD/S-LRA Model

Microscopic models that evaluate the absolute electrostatic energy (including the PDLD model) do not involve any dielectric constant. This provides a rigorous formulation and prevents one from falling into various traps, but it is also sometimes a practical 'weakness' since they involve the balance between very large contributions, which is hard to obtain without the elusive ability to obtain a perfect convergence. In some cases, it is possible to assume that the balance must exist and to represent this balance by scaling the results by a dielectric constant. Such a philosophy leads to semi-macroscopic models, which are obviously more stable (but not necessarily more reliable) than the corresponding microscopic models. A practical way of exploiting the stability of the semi-macroscopic models, while keeping a clear physical picture, is the semi-macroscopic version of the PDLD model: the PDLD/S model[91,92]. This model starts with the PDLD model and then scales the corresponding contributions by assuming that the protein has a dielectric constant ε_p. Furthermore, in order to reduce the unknown factors in ε_p the model transforms the PDLD/S model to the PDLD/S-LRA model, where the PDLD/S energy is evaluated within the LRA approximation with an average on the configurations generated by molecular dynamics simulation of the charged and uncharged states (see Refs. 67,73,93). In this way one uses molecular dynamics simulations to automatically generate protein configurations for the charged and uncharged forms of the given "solute" and then average these contributions according to Eq. (III.2). This treatment expresses the free energy of moving a charge from water to the protein active site, which is given by:

$$\Delta\Delta G_{sol}^{w\to p} = \frac{1}{2}\Big[\left\langle \Delta U^{w\to p}\left(Q=0 \to Q=Q_0\right)\right\rangle_{Q_0} + \left\langle \Delta U^{w\to p}\left(Q=0 \to Q=Q_0\right)\right\rangle_{Q=0}\Big] \quad \text{(III.5)}$$

where

$$\Delta U^{w\to p} = \left[\Delta\Delta G_p^w\left(Q=0 \to Q=Q_0\right) - \Delta G_Q^w\right] \times \left(\frac{1}{\varepsilon_p} - \frac{1}{\varepsilon_w}\right) + \Delta U_{Q\mu}^p\left(Q=Q_0\right)\frac{1}{\varepsilon_p} \quad \text{(III.6)}$$

where ΔG_Q^w is the solvation energy of the given charge in water, $\Delta\Delta G_p^w$ is the change in the solvation energy of the entire protein (plus its bound charge) upon changing this charge from its actual value (Q_0) to zero, and $\Delta U_{Q\mu}$ is the microscopic electrostatic interaction between the charge and its surrounding protein polar groups (for simplicity we consider the case when the interaction with the protein ionizable groups is treated in a separate cycle (e.g., Ref. 73) with the effective dielectric constant ε_{eff}). Now we bring here the explicit expression of Eq. (III.6), since to the best of our knowledge it provides the clearest connection between microscopic quantities such as $\Delta\Delta G_p^w$ and $\Delta U_{Q\mu}$ and semi-macroscopic treatments. In fact, the PDLD/S treatment without the LRA treatment converges to the PB treatment and thus provides the clearest connection between PB treatments and microscopic treatments, allowing one to demonstrate the exact nature of ε_p (e.g., see Ref. 93 and Section IV.4). At any rate, it is important to keep in mind that the LRA considerations treat the protein reorganization explicitly, and this reduces the uncertainties about the magnitude of ε_p. It is also useful to note that the so-called molecular mechanics Poisson-Boltzmann surface area (MM-PBSA) model[94] is apparently an adaptation of the PDLD/S-LRA idea of molecular dynamics generation of configurations for implicit solvent calculations, while only calculating the average over the configurations generated with the charged solute (the first term of Eq. (III.2)).

III.4 Macroscopic Continuum Approaches Depend Critically on the Assumed ε_p

As stated in Section III.2, early macroscopic continuum models (e.g., Ref. 95) and DC models, or PB models[96–98], have treated the protein as a uniform dielectric medium without considering the effect of the protein's permanent dipoles ($\Delta U_{Q\mu}$ in Eq. (III.6)). The Tanford-Kirkwood model[95] also omitted the self-energy term (see discussion in Refs. 76, 99). However, more recent PB models resemble the PDLD model in considering the shape of the protein in detail and including the protein or solvent permanent dipoles explicitly, while using a macroscopic-continuum approach to treat dielectric effects[100–113]. These models, as implemented in the programs DelPhi[97], UHBDP[114] and APBS[115] map the protein-solvent system to a cubic grid with two or more dielectric regions. The region representing the solvent usually is assigned a macroscopic dielectric constant similar to that of water ($\varepsilon_s \approx 80$), while that representing the protein is given a lower dielectric constant (ε_p) between 2 and 40. The electric potential $U(r)$ at each grid point is evaluated by solving the PB equation subject to the boundary condition that the potential goes to the value given by the Debye–Hückel theory at long distances from the protein. As stated above, the nature of the PB results can be best analyzed by using the PDLD/S formulation. With this in mind, the electrostatic energy of a single charge Q in processes such as reduction or protonation can be written as

$$G_{elec} = \Delta U_{Q\mu}/\varepsilon_p + G_{RF} \tag{III.7}$$

where $\Delta U_{Q\mu}$ is again given by Eq. (III.6) and G_{RF} (the "reaction field" energy) represents interactions of the protein charges with the fields from the solvent and electrolytes. At low ionic strengths, the PB equation can be linearized by making the approximation, $\sinh[\varepsilon(\mathbf{r})] \approx \varepsilon(\mathbf{r})$.

As mentioned above, the Tanford-Kirkwood model and other early macroscopic models had major inconsistencies because they ignored the microscopic nature of the protein's permanent dipoles. Interestingly, the almost complete neglect of the self-energy contribution of the permanent dipoles by the continuum workers was subsequently introduced by calling this missing contributions "back field", perhaps in order to hide its original omission. At any rate, current PB models represent these dipoles explicitly, and thus are semi-macroscopic models[116].

Some DC implementations also include sampling of multiple protonation states and rotational conformers of polar hydrogen bond donors[117–120] which, in many respects, resembles the earlier and more rigorous LRA treatment in the PDLD/S-LRA model described above.

Regardless of the above developments, it is important to remind the readers that the popularity of continuum models is largely due to the belief that the formulation of the PB equation must reflect some fundamental physics and that the criticism of such models only addresses minor points that can be 'fixed' in one way or another. The discussion of conceptually trivial issues such as the linear approximation of the second term in Eq. (II.6) frequently created the sense that this is the true problem in such models. In fact, the real unsolved problem is the fact that the results depend critically on ε_p, whose value cannot be determined uniquely. Another reason for the acceptance of PB models has been the use of improper validation, addressing mainly trivial cases of surface groups where all macroscopic models would give the same results (see Section IV.5). At any rate, PB models (e.g. Ref. 120) are becoming more and more microscopic, reflecting ideas from the semi-macroscopic PDLD/S-LRA model.

III.5 The Generalized Born Model is Simply a Screened Charge-Charge Interaction Model

In treating electrostatic energies in macromolecules, it is convenient to use a two-step thermodynamic cycle. The first step involves the transfer of each ionized group from water to its specific protein site where all other groups are neutral, and a second step where the interaction between the ionized groups is turned on[93]. Using this cycle, we can write the energy of moving ionized groups from water to their protein site as:

$$\Delta G = \sum_{i=1}^{N} \Delta G_i + \sum_{i>j} \Delta G_{ij} \qquad \text{(III.8)}$$

where the first and second terms correspond to the first and second steps, respectively. ΔG_i, which represents the change in self-energies of the ith ionized group upon transfer from water to the protein site, can be evaluated by the PDLD/S-LRA approach with a relatively small ε_p. However, the charge-charge interaction terms (the ΔG_{ij}) are best reproduced by using a simple macroscopic Coulomb-type law and by expressing the overall electrostatic energy by

$$G_{elec} = G_{sol}^{\infty} + 332 \sum_{i=1}^{n-1} \sum_{j=i+1}^{n} Q_i Q_j / r_{ij} \varepsilon_i \tag{III.9}$$

where G_{sol}^{∞} is the sum of the solvation free energies of the individual charges at infinite dilution. The dielectric constant ε_{ij} can be described by (e.g., Ref. 99):

$$\varepsilon(r_{ij}) = 1 + \varepsilon'[1 - exp(-\mu r_{ij})] \tag{III.10}$$

where ′ is 60 and ε' is in the range of 0.1 to 0.2. Although the fact that ε_{ij} is large has been supported by mutation experiments (e.g., Refs. 121, 122) and conceptual considerations[52], the use of Eq. (III.10) is still considered by many as a poor approximation for PB treatments (without realizing that the PB approach depends entirely on an unknown ε_p). Interestingly, the so-called generalized Born (GB) model, whose usefulness is now widely appreciated[123], is basically a version of Eq. (III.9).

With

$$G_{sol(i)}^{\infty} = -332\left(1 - 1/e_{eff}\right)\left(Q_i^2 / 2a_i\right)$$

this point is systematically analyzed in Ref. 124. Thus, the GB model is merely a version of an earlier treatment[52].

Overall we conclude that the widely accepted GB treatment is in fact a glorified Coulomb's law treatment[116]. Nevertheless, it is instructive to note that the GB approximation can also be obtained by assuming a local model where the vacuum electric field, $\mathbf{E}_0$, and the displacement vector, $\mathbf{D}$, are identical[107]. This model, which can be considered as a "local Langevin dipole model"[52], or as a version of the non-iterative LD model, allows one to approximate the energy of a collection of charges and to obtain (with some assumptions about the position dependence of the dielectric constant) the GB approximation.

It should be noted that the simple derivations of the GB approximation[52] are valid for homogeneous media with a high dielectric (e.g., the solvent around the protein). This can be problematic when one considers models where the protein has a low dielectric constant and when we are dealing with transfer of charges from water to the protein environment. Fortunately, it is reasonable to use a large ε in the second term in Eq. (III.9), even for protein interiors.

Instructively, we started to use very early the relationship that is now called GB. This was done once we realized that this relationship is a very powerful way of estimating the electrostatic energy of reacting fragments in reverence reactions in solution. This gave us a major advantage of being able to estimate the reverence reactions to enzymatic reactions. An example is given in Eq. (20) of Ref. 125 that was used to estimate the energetics of the reference reaction of chymotrypsin.

Regardless of the effectiveness of the GB treatment we must emphasize that it hides the true origin of the solvation energy at a specific distance (namely the interaction with the surrounding solvent at the given site, rather than a screening of electric fields). This point is related in some ways to the fact that key workers, in the field of modeling biological functions, ignored until a rather late stage the crucial self-energy contribution (e.g., Refs. 126 and 300).

III.6 A Short Summary of the Hierarchy of Electrostatic Models

Although the above sections give a rather comprehensive description of the main electrostatic models, it is useful to summarize here the key hierarchy of the electrostatic models. This hierarchy is summarized in Fig. III.4. Basically, consistent models of electrostatic energies in proteins range from microscopic all-atom models that describe explicitly the protein and the surroundings solvent, to the PDLD model and the PDLD/S model. To this list we should add the DC models (regardless of the limitations of such models).

At this point it might be interesting to add some personal perspective on the LD model. In 1975, I was asked by David Blow to give a short lecture on this water model in a Faraday discussion meeting about water. I found out that I was the only speaker who talked about solvation[127] and unfortunately no one appeared to follow my inverted logic of using the solvation to calibrate the polarization of the Langevin dipoles (rather than evaluating the solvation by a first principles model). The fact that the water system was described by a simple dipole-based model led subsequently to never-ending criticism. For example, it was argued that it is incorrect to use dipoles when everyone knows that a water molecule is not a dipole (this argument missed, of course, the fact that our dipoles were "effective" and were calibrated based on solvation free energies). Actually, for about ten years, we were arguably the only group with a clear physical understanding of solvation and

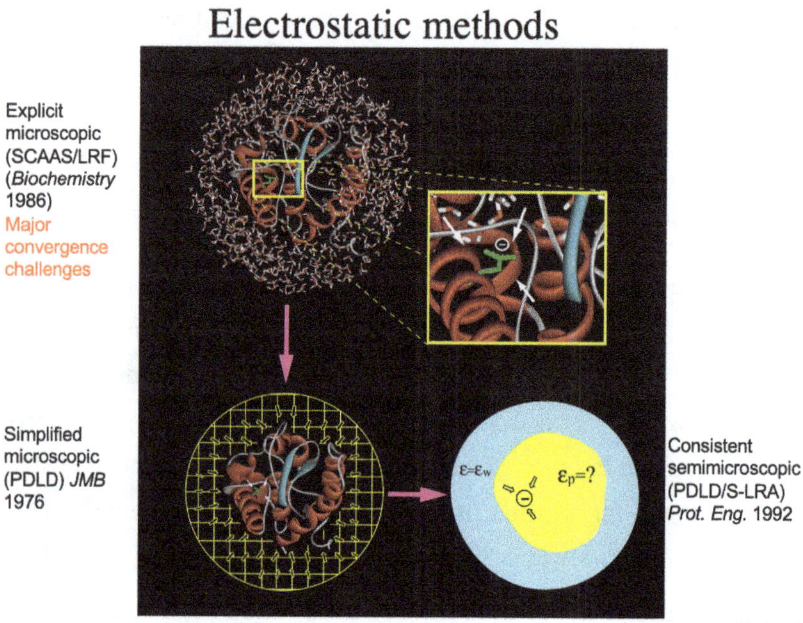

Fig. III.4 Hierarchy of electrostatic models. Moving from a full microscopic model that includes explicitly the protein permanent and induced dipoles as well as the surrounding water molecules, to the PDLD model that includes explicitly the protein permanent and induced dipoles and represents the solvent by Langevin dipoles, and the PDLD/S model which scales the PDLD model in a consistent way by a protein dielectric. Reproduced with permission[93].

electrostatics in proteins. This was the case since we insisted on using a simple but complete model for studying the solvation of charges in proteins, while the rest of the community was satisfied with either continuum models that used an undefined dielectric for the protein, or with calculations that considered a single ion or a single HCl molecule as a guide for studying solvation effects.

Chapter IV

Modeling Charges in Proteins

Early studies of electrostatics in proteins had major problems with their physical foundation. These studies considered the protein as a homogeneous low dielectric medium which is clearly unjustified for heterogeneous environments. Despite this obvious difficulty, the early studies of electrostatic effects in proteins have been marked by the emergence of phenomenological models[14,95]. Perhaps the most influential model was the abovementioned Tanford-Kirkwood (TK) model[95] that consider the protein as a sphere with a uniform dielectric constant. While the TK model has helped in guiding an inspiring early study of electrostatic interactions in proteins, it reflected assumptions that were formulated before the emergence of actual protein structures. For example, it was assumed that the ionized groups are located on the protein surface. This assumption was found to be incorrect in key cases. Subsequent analysis by Warshel and co-workers[99] indicated that the model ignores one of the most important physical aspects of protein electrostatics: the self-energy of the charged groups (the energy of forming these charges in the given protein site). Nevertheless, the TK model is still invoked as a physically reasonable model, where only the assumption of a spherical shape is considered as a rough approximation.

The work of Warshel and Levitt[69] marked the emergence of modern studies of electrostatic effects in proteins. This work started from the realization that the only way to reach unique conclusions about the energetics of enzyme catalysis was to abandon the concept of a dielectric constant and to develop a simplified microscopic model that included all feasible microscopic contributions. The solvent was modeled by using an explicit grid of Langevin dipoles and the protein was modeled explicitly, taking into account its permanent and induced dipoles, as well as a limited reorganization (by energy minimization). The resulting protein dipoles Langevin dipoles (PDLD) model has been the first model to correctly capture the physics of protein electrostatics. In fact, the microscopic perspective of the PDLD model has been the best guarantee against the traps of the continuum concepts (see below).

The introduction of discretized continuum (DC) models[96,108,128] (see also Chapter III) as well as the enormous impact of the authoritative continuum formulation (e.g., Eqs. II.4–II.6) led to a wide acceptance of DC models. This trend was not reduced by the identification of major conceptual problems that were pointed out and resolved by the use of microscopic models[52,99,109,116]. For example, early DC models overlooked self-energy, ignored the key role of the protein permanent dipoles[129] and used unjustified dielectric constants[93,130]. However, the continuum models kept evolving and becoming more and more microscopic and in many respects converging to the PDLD and its semi-macroscopic version, the PDLD/S-LRA model.

At present we have a wide range of reasonable electrostatic models, but there are still major misunderstandings about the nature and meaning of different aspects of these models (e.g., the dielectric constant). Some of these issues that were considered will be clarified in Section IV.4 and elsewhere. At any rate, it is important to keep in mind that almost every conceptual problem of the continuum models has been resolved and clarified using microscopic concepts. Perhaps the best example is the elucidation of the nature of the protein dielectric constant (that will be discussed below).

From historical perspective it is useful to mention that insightful observations of Perutz[130] and others (e.g., Refs. 110–112) have indicated that electrostatic interactions must play an important role in biology. However, it was the development of computational methods and their successful applications that led to the realization that electrostatic energies provide the best structure-function correlation for proteins[129]. This has been demonstrated in studies of a wide range of properties that will be considered in subsequent sections.

IV.1 pK_as of Ionizable Residues in Proteins

The charges of ionized protein residues play a crucial role in various biological processes including enzymatic reactions, proton transport, and protein stability. Thus, the ability to evaluate the energetics of the ionized residues and the corresponding pK_a groups are crucial in attempts to correlate the structure and function of proteins. This is also highly important in in any attempts to validate different models for electrostatic energies in proteins[52].

All-atom pK_a calculations by free energy perturbation (FEP) approaches started with the work of Ref. 72 (see Fig. IV.1). More recent works include studies of the pK_a of metal-bound water molecules[113] and proton transfer in proteins[131] as well as functionally important groups (e.g., Refs. 132–134). Such calculations are in principle reliable but they depend of course on the convergence.

All-atom linear response approximation (LRA) calculations[73,74] have become useful and in many cases they provide an excellent approximation for the FEP results. Attempts to estimate the error range in the calculations (e.g., Ref. 64) determine that the error range of the all-atom models is still problematic, although the inclusion of proper long-range

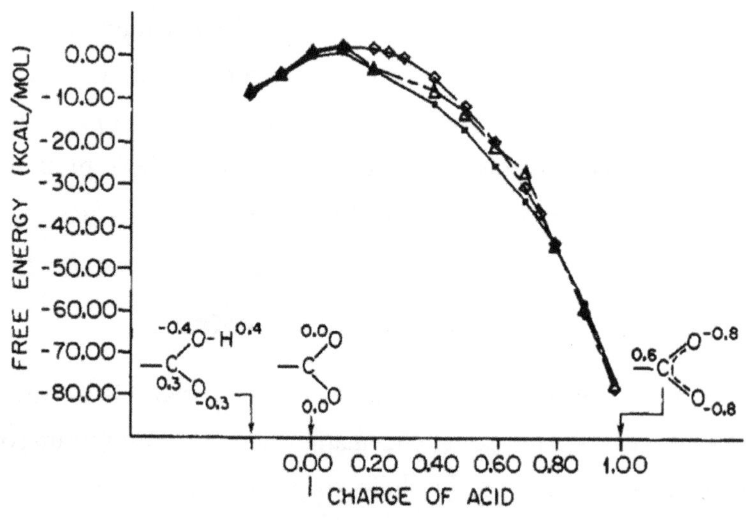

Fig. IV.1 Calculated free energy of ionization for Asp-3 (Δ) and Glu-7 (■) in bovine pancreatic trypsin inhibitor and aspartic acid (◇) in solution. Reproduced with permission[72].

treatments and induced dipoles leads to some improvements[73,76]. Of course, efficient sampling of the protein configurations is crucial. pK_a calculations by the semi-macroscopic PDLD/S-LRA model[73,93] usually yield encouraging results with $\varepsilon_p = 4$. In such calculations the LRA treatment accounts in part for the protein reorganization. However, the limit of $\varepsilon_p = 2$ (which would correspond to a 'perfect' model with implicit induced dipoles) is not reached, since some of the reorganization and water penetration is not reproduced in a reasonable simulation time.

Semi-macroscopic "continuum" (i.e., Poisson-Boltzmann (PB)) models, which at some late stage started to include the protein permanent dipoles and the self-energy term, also sometimes give reasonable results. Yet most PB models[135–137] do not consider explicitly the effect of the protein reorganization, and even the so-called multi-conformation continuum electrostatic model[120] (which has been inspired by the LRA approach) considers the reorganization of polar side chains but does not perform a proper LRA treatment on all the protein coordinates. Furthermore, current PB models use the same ε_p for treating the effect of the protein dipoles and for providing the screening of charge-charge interactions. The corresponding inconsistencies cannot be demonstrated by the use of discriminating benchmarks (see Section IV.3) that consider both self-energies and charge-charge interactions[138].

Many workers (e.g., Refs. 139–141) presumed that the problems with PB calculations of pK_as reflect the need for averaging over the protein configurations. In fact, configurational averaging involves only one charge state. This corresponds to only one charge state in the LRA treatment and thus to incorrect estimates of the relevant charging free energy.

It is likely that converging microscopic pK_a calculations will lead eventually to stable quantitative results. At such a stage a comparison of the microscopic and semi-macroscopic models (as was done in Ref. 25) will be very useful.

One of the best examples of the convergence problem associated with pK_a calculations is the challenge of evaluating the pK_a of Glu 66 in the hydrophobic site of staphylococcal nuclease[142,143] and related cases. Accounting for the observed effect has been very challenging for seemingly rigorous microscopic models. Our standard FEP calculations gave major deviations of 23 pK_a units from the observed pK_a of Glu 66 (a shift of about 5 units), despite running simulations of 5 ns. This result indicates that we are missing local unfolding and/or water penetration processes upon ionization of Glu 66. The challenge is to capture this configurational change by the available simulation times. Our strategy for this case involved the "overcharging approach". In this approach we induce the conformational change by introducing a charging coordinate where we decrease the charge from $Q = -1.0$ to $Q = Q_i$ where $|Q_i| > 1.0$. Performing such a charging and unchanging cycle allows us to overcome the barrier for the local unfolding and water penetration and exposure of the ionized Glu residue to the solvent. The overcharging approach[144] that is described in Fig. IV.2 allows us to explore the free energy landscape of buried ionized groups and to obtain the relevant pK_as in a reasonable computer time.

At this point it may be useful to mention my own perspective on pK_a calculations. In 1975, while working on the calculations of the energetics of the catalytic reaction of lysozymes[10], I concluded that I must understand the thermodynamic cycle of the ionization of aspartic

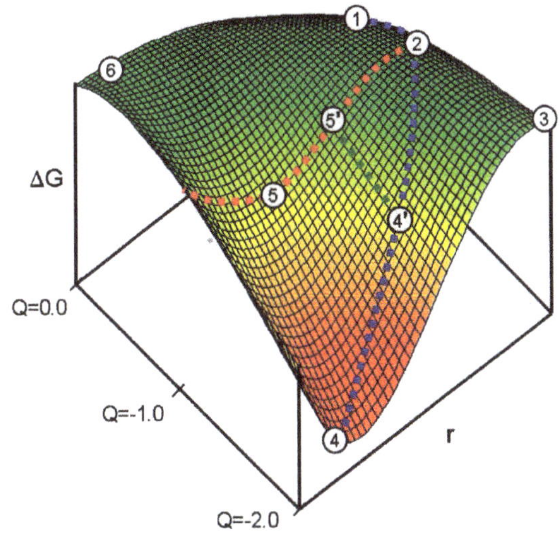

Fig. IV.2 A schematic description of the overcharging procedure. The figure describes the free energy surface as a function of the ionizable residue and an effective protein coordinate, **r**. Our mapping procedure is aimed at evaluating $\Delta G_{1\to5}$ and this is accomplished by calculating $\Delta G_{1\to5'} = \Delta G_{1\to4'} + \Delta G_{4'\to5'}$ and then finding the lowest $\Delta G_{1\to5'}$. Reproduced with permission[144].

acid in the gas phase, in solution and in the protein. This study led me to the realization that I am dealing with the balance between very large energy contributions and thus have to generate a stable model. On the other hand, the available continuum models of proteins did not provide any indication of the very large self-energy and the contribution to this term from different factors in the protein, e.g., the water molecule in and around the protein, plus the protein permanent and induced dipoles. Evaluating these terms appeared to provide a powerful insight. Interestingly, in a meeting in Italy in 1988 I was criticized by a distinguished scientist, who asked "why do you have to calculate the absolute solvation energy when you study the pK_a shift in different sites of a protein?" I said something about insight but he dismissed it.

However, in 1990 he already published in *Science* a paper[145] on the same type of calculations. At any rate, the insistence on full microscopic contributions to the solvation free energy have kept me and my coworkers from many conceptual mistakes in assessing the energetics of charges in proteins.

IV.2 Ion Pairs

Ion pairs or "salt bridges" play a major role in biological systems and biological processes. This includes ion pairs that help in maintaining protein stability, ion pairs that play crucial roles in signal transduction, and ion pairs that play a major role in photobiological processes (e.g., Ref. 146), as well as the salt bridges that control the Bohr effect in hemoglobin[147] and ion pair-type transition states[18]. The analysis of the energetics of ion pairs in proteins has been the subject of major conceptual misunderstandings (see discussions in Refs. 52, 99). The most problematic confusion is the presumption that ion pairs are more stable rather than less stable in nonpolar environments. This point is demonstrated in Fig. IV.2 (see also the analyses in Refs. 148, 149).

Examples of the related confusion are provided in Ref. 150 and in most of the desolvation models (e.g., Refs. 151–153) where it has been assumed that enzymes destabilize their ground states by placing them in nonpolar regions. However, enzymes do not provide nonpolar active sites for ion pair-type ground states (they provide a very polar environment that stabilizes the transition state as discussed in, e.g., Refs. 18, 154). Since ion pairs are not stable in nonpolar sites in proteins, most ion pairs in proteins are located on the surface. However, key ion pairs, which are located in protein interiors, are always surrounded by polar environments[18]. Forcing ion pairs between amino acids to be in nonpolar environments usually results in the conversion of the pair to its nonpolar form (by a proton transfer from the base to the acid).

The energetics of specific ion pairs in proteins was studied by microscopic FEP calculations[155,156] and semi-macroscopic calculations[84,99,157–159]. However, we have relatively few systematic validation studies. Obviously, it is essential to have more validation studies considering internal ion pairs with well-defined structure and energy. In this respect, it is useful to point out that the statistical study of Ref. 160 is likely to involve similar problems to those that will be mentioned in Section IV.6, since it is not based on a discriminative database; it includes mainly surface groups rather than ion pairs in protein interiors.

The importance of consistent analysis of ion pair energetics has been provided by Schutz and Warshel[93] who pointed out major problems in the PB analysis of Ref. 161 that used a low value of ε_p. This led to the problematic prediction of major protein destabilization by ion pairs[161]. As will be discussed in Chapter IX, a careful analysis of ion pair energetics indicates that such pairs contribute in many cases to protein stability. The problem in the analysis of PB approaches with low ε_p is that they do not allow polar groups to reorient in response to the charges of the ion pair. It is exactly the stabilizing effect of the protein polar groups that plays a crucial role in stabilizing ion pairs in proteins. Clearly, the estimate of the energetics of ion pairs depends crucially on the dielectric constant used and models that do not consider the protein relaxation explicitly must use larger ε_p than the PDLD/S-LRA model.

At the end of this section, it is useful to elaborate on my own perspective on the energetics of ion pairs. Naturally it is very tempting to think of the energetics of ion pairs in terms of Coulomb's law with electrostatic attraction scaled by a dielectric constant. However, as clarified above, this treatment does not provide the energy of forming the charge in the given environment and the effect of change in the environment around the charges. This led to a major confusion in the field including even the great work of Tanford and Kirkwood (see above). One of the best examples occurred at a conference on protein folding at Santa Cruz in 1984, where I discussed my work on ion pairs and Martin Karplus intercepted me, saying "but ion pairs are very stable in the gas phase in drops of organic molecules." Of course, he missed the point shown in Fig. IV.3. That is, in the gas phase the nonpolar drop is surrounded by vacuum, but in solution the drop will explode and the ions will go to solution from the low dielectric environment. In order to drive my concept forward I published a paper in *Nature*[155] on the inherent problem in ion pair reversal in proteins, promising a financial prize to anyone who succeeds in creating a more stable ion pair in protein by inverting the original pair. This work helped in pointing out that the energetics of ion pairs are determined by their specific environment (which should be complementary to the original ion pair).

Gradually some of the workers in the community started to change their story and moved from the idea of ion pairs in low dielectric (namely a nonpolar environment) to the argument that "yes, it is nonpolar environments but with hydrogen bonds." Of course, hydrogen bonds are polar rather than nonpolar environments.

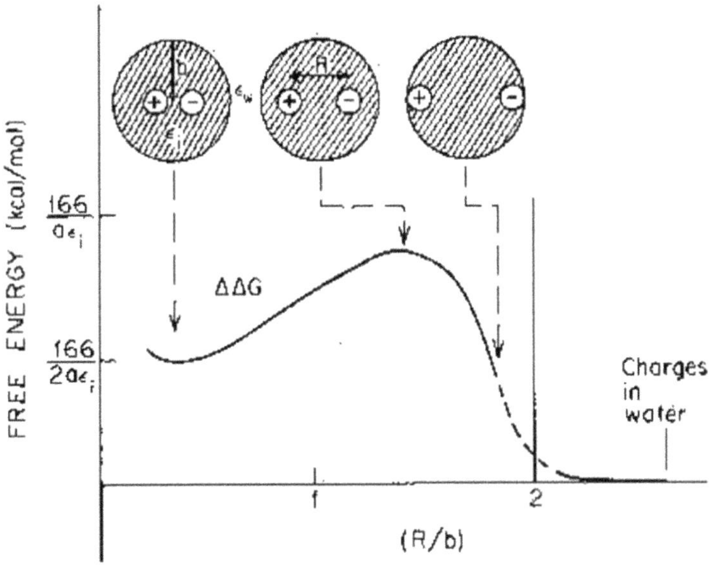

Fig. IV.3 Electrostatic free energy of an ion pair inside a nonpolar sphere of dielectric ϵ_i surrounded by a continuum with the dielectric constant of water. The figure demonstrates that ion pairs prefer to be in polar environments. Reproduced with permission[350].

It is also useful to point out that the traditional picture of ion pairs in a dielectric environment involves the concept of a very large opposite field from the dielectric medium that balances the attraction between the pair. However, my picture of a balance of the charge-charge interaction energy by the change in solvation energy (e.g., Fig. IV.5) appeared to be a much more useful and physically clear picture (see discussion in Section IV.4).

It is important to note that ion pair interaction also contributes in a major way to protein folding energy, although continuum estimates predicted the opposite trend (see discussion in Chapter IX).

Finally, we note that ion pair-type transition states play a major role in enzymatic reactions and the stabilization of such ion pairs by the preorganized protein environment plays a major role in enzyme catalysis (see Chapter V).

IV.3 The Macrodipole is a Problematic Physical Description

The macroscopic dipoles of alpha helices had created the impression that such helices contribute inside proteins in a similar way to how they contribute in the gas phase[162,163]. The main reason for the acceptance of the helix dipole idea (except the structural appeal of this proposal) has been the use of incorrect dielectric concepts. The problematic estimates of large helix dipole contribution[163–167] reflect a drastic underestimate of the dielectric constant and the customary tendency to avoid proper validation studies. Basically, most estimates of the magnitude of the helix dipole effect have not involved any serious verification by using the same model in evaluation of observables such as pK_a shift and enzyme catalysis. Our quantitative estimate of the effect of the helix dipole[168] established that the actual effect is due to the first few microscopic dipoles at the end of the helix and not to the helix macrodipole. It was also predicted that neutralizing the end of the helix by an opposing charge will have a very small effect and this prediction was confirmed experimentally[169].

The need for proper considerations of the helix dipole effect has been demonstrated in the case of the KcsA K^+ channel. The pioneering PB study of Roux and MacKinnon[170] used $\varepsilon_p = 2$ and obtained an extremely large effect of the helix dipoles on the stabilization of the K^+ ion in the central cavity (~–20 kcal/mol). However, a much more consistent PDLD/S-LRA study[171] gave a much smaller effect of the helix macrodipole. Basically, the use of $\varepsilon_p = 2$ overestimates the effect of the helix dipole by a factor of 3 and the effect is rather localized on the first few residues. Another example that will be discussed in Section XI.2 is the problematic analysis of the helix dipole in aquaporin where it has been suggested that the barrier for proton transport is due to the helix dipole[172,173]. However, the corresponding careful analysis of Ref. 174 established that we are dealing with a major overestimate.

IV.4 What is the Protein Dielectric Constant?

It is tempting to assume that the dielectric constant of proteins is a quantity that can be used in a universal way in different parts of the protein and in different models. In fact, it has been assumed for a long time that proteins can be represented by low dielectric (2–4), and this assumption has been supported by some microscopic simulations (see below). However, a careful microscopic-based conceptual and practical analysis[47,58] indicated that the value of the dielectric constant of proteins is entirely dependent on the method used to define this constant and on the model used in its evaluation. This view might look strange to those readers who are accustomed to the view that macroscopic models have a unique meaning. It might also look puzzling to workers who are familiar with microscopic statistical mechanics, where the fluctuations of the total dipole moment of the system determine its dielectric constant in a unique way. However, (see discussion in Refs. 93,175,176) the protein dielectric constant ε_p has little to do with what is usually considered as the protein dielectric constant. This fact can be best realized by considering different limiting cases. That is, when all the interactions are treated explicitly, $\varepsilon = 1$; when all but induced dipoles are included explicitly, $\varepsilon = 2$; and when the solvent is not included explicitly, $\varepsilon > 40$ (although this is a very bad model). When the protein permanent dipoles are included explicitly but their relaxation (the protein reorganization) and the protein-induced dipoles are included implicitly, the value of ε_p is not a well-defined quantity. In this case, ε_p should be between 4 and 6 for dipole-charge interactions and 4–10 for charge-charge interactions[177,178].

It was popular to assume that there is only one "proper" dielectric constant ($\varepsilon = \bar{\varepsilon}$), which is presumably the one obtained from the fluctuation of the average dipole moment[176,179–182]. This assumption has been subsequently replaced by the assumption that the finding of a high effective dielectric constant in proteins reflects the effect of ionized surface groups on the simulated $\bar{\varepsilon}$ (see below). However, the dielectric constant in protein interiors is not a constant because it depends on the site considered[156,160]. More importantly, as discussed above $\bar{\varepsilon}$ cannot be equal to ε_p, where the simplest example is the use of explicit models where $\varepsilon_p = 1$. Detailed analysis of this issue is given in Ref. 183. At any rate, it is still useful to review what has been learned on the nature of $\bar{\varepsilon}$.

Arguably the most consistent analysis of the macroscopic protein dielectric constant ($\bar{\varepsilon}$) has been performed in Ref. 176 where it was found that the value of $\bar{\varepsilon}$ can be significantly larger than commonly assumed. This work calculated the dielectric not by the widely used fluctuation formula, but by the much more fundamental formula

$$(\bar{\epsilon} - 1)E = 4\pi P \tag{IV.1}$$

where E is evaluated by Eq. (II.17).

The corresponding results for different regions of trypsin are given in Fig. IV.4.

It was found that $\bar{\varepsilon}$ is larger than the traditional value ($\bar{\varepsilon} \cong 4$) assumed based on measurements of dry proteins and peptide powders[184,185], or from simulations of the entire protein, rather than a specific region (e.g., Ref. 176). In fact, the early finding of $\bar{\varepsilon} = 4$, which was obtained from a gas phase study[185], ignored the effect of the solvent reaction field (see analysis and discussion in Ref. 176). The larger than expected value of $\bar{\varepsilon}$ was attributed to the fluctuations of ionized side chains[122,160,167]. However, the consistent inclusion of the reaction field of the solvent[176] gave $\bar{\varepsilon} \cong 10$ in and near protein active sites, in the absence of fluctuations of the ionized residues. Thus we established that the fluctuations of the protein polar groups and internal water molecules contribute significantly to the increase in $\bar{\varepsilon}$. We also note experimental studies[186,187] that provided a clear support to the finding that $\bar{\varepsilon}$ is large in active sites even in regions that are distanced from the protein surface.

It should be clarified that $\bar{\varepsilon}$ is not so relevant to electrostatic energies in proteins. Here one should focus on the energies of charges in their actual environment and the best ways for a reliable and effective evaluation of these energies. In this respect it is important to recognize that $\bar{\varepsilon}$ does not tell us in a unique way how to determine the ε_p of semi-macroscopic models. For example (see also above), obtaining $\varepsilon_p = 1$ in a fully microscopic model, or $\varepsilon_p = 2$ in a model with implicit induced dipoles, has nothing to do with the corresponding $\bar{\varepsilon}$. The best way to realize this point is to think of charges in a small water sphere (our hypothetical

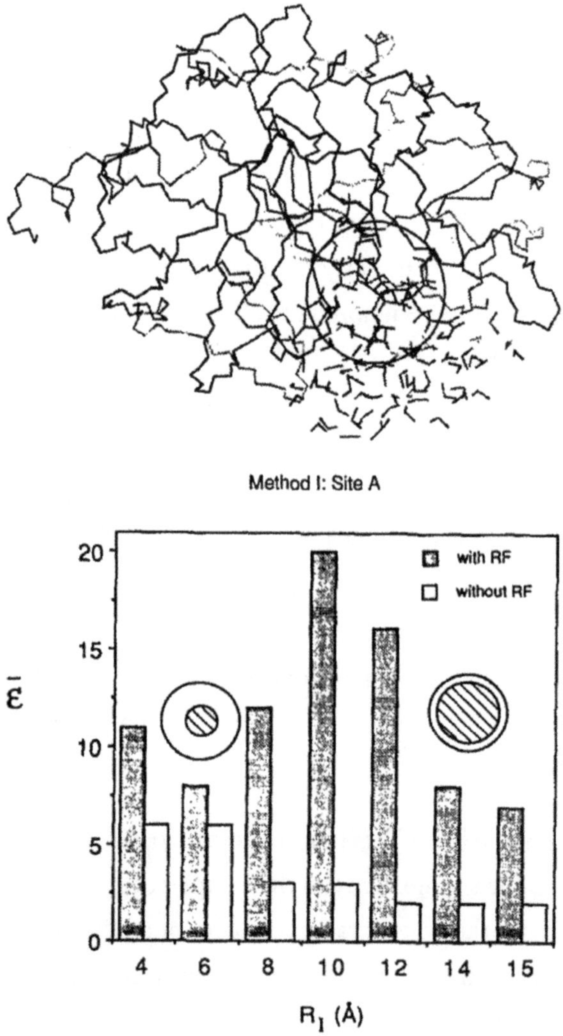

Fig. IV.4 Effect of the solvent on $\bar{\epsilon}$ for site A. The upper diagram shows the trypsin system and the corresponding region (I) for site A (the oxyanion hole). The trypsin molecule is represented by its main-chain trace and the water molecules are plotted explicitly. The circle contains the heavy atoms found within 4 Å from site A and is used to represent the case where region (I) is a 4 Å sphere. The lower diagram gives $\bar{\epsilon}$ as a function of the radius of region (I), though only two circles representing R_I of 6 and 14 Å are shown. The calculations were done using method I with a simulation time of 50 ps. Reproduced with permission[176].

protein) surrounded by water. In such a model, $\bar{\varepsilon} \cong 80$, whereas microscopic models of the same system with implicit induced dipoles will be best described by using $\varepsilon_p = 2$. Of course, one may argue that the entire concept of a dielectric constant is invalid in the heterogeneous environment of proteins[57]. However, since fully microscopic models still do not give sufficiently precise results, it is very useful to have reasonable estimates using implicit models and, in particular, semi-macroscopic models. Thus, it is justified to look for optimal ε_p values after realizing what these parameters really mean. The use of a consistent semi-macroscopic model, such as the PDLD/S-LRA, does not guarantee that the corresponding ε_p will approach a universal value. For example, in principle the PDLD/S-LRA model should involve $\varepsilon_p = 2$, because all effects except the effect of the protein induced dipoles are considered explicitly. However, the configurational sampling by the LRA approach is not perfect. The protein reorganization energy is probably captured to a reasonable extent[163], but the change in water penetration on ionization of charged residues is probably not reproduced in an accurate way. This problem is particularly serious when we have ionized groups in nonpolar sites in the interior of the proteins. In such cases we have significant changes in water penetration during the ionization processes (as demonstrated in the very instructive study of Ref. 188) but these changes are unlikely to be reproduced in nanosecond simulation times.

The important observation[99,189,190] that the dielectric constant for charge-charge interactions, ε_{eff}, is larger than what is usually assumed was rarely analyzed with a clear microscopic perspective. For example, Mehler and coworkers correctly argued[138,190] that $\varepsilon_{eff}(r)$ in polar solvents is a sigmoid function with a large value at relatively short distances based on classical studies of $\varepsilon(r)$ in water[191–193]. Similarly, it has been assumed by Jonsson and coworkers[189] that the electrostatic interactions in proteins can be described by using a large ε_{eff} for charge-charge interactions. Unfortunately, although we repeatedly advanced the view that ε_{eff} is large in many cases[52], we believe that the reasons for this are much more subtle. That is, the finding of a particular behavior of ε_{eff} in water might be not so relevant to the corresponding behavior in proteins (in principle, ε_{eff} in proteins can be very different from ε_{eff} in water). Here the remarkable reason ε_{eff} is large, even in proteins, was rationalized in Ref. 52, where it was pointed out that the large ε_{eff} must reflect the compensation of charge-charge interactions and solvation energy in proteins (see Fig. IV.5). It seems to us that the understanding of this

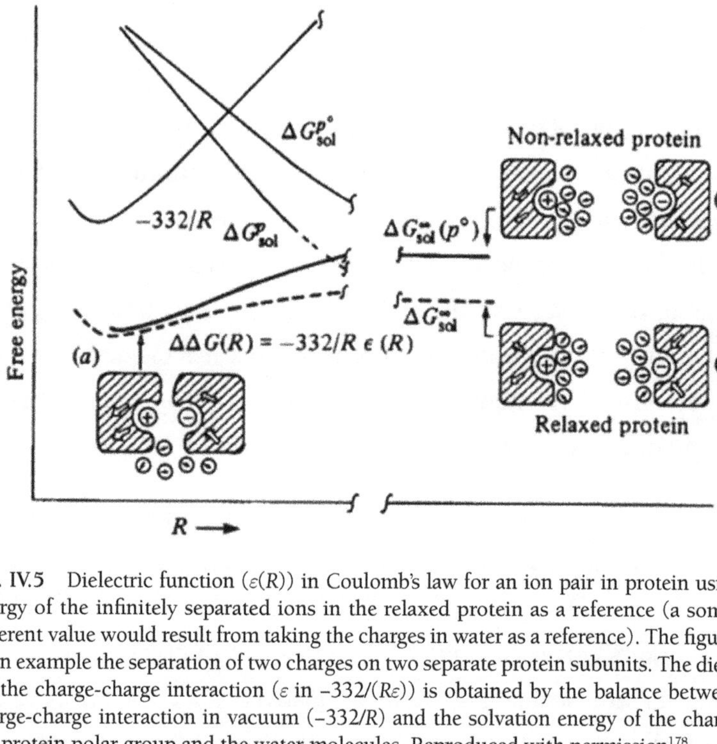

Fig. IV.5 Dielectric function ($\varepsilon(R)$) in Coulomb's law for an ion pair in protein using the energy of the infinitely separated ions in the relaxed protein as a reference (a somewhat different value would result from taking the charges in water as a reference). The figure uses as an example the separation of two charges on two separate protein subunits. The dielectric for the charge-charge interaction (ε in $-332/(R\varepsilon)$) is obtained by the balance between the charge-charge interaction in vacuum ($-332/R$) and the solvation energy of the charges by the protein polar group and the water molecules. Reproduced with permission[178].

crucial compensation effect is essential for comprehending the nature of ε_{eff} in proteins.

IV.5 Electric Fields

The approach described in Section II.1 allows one to obtain the local macroscopic field proteins by integrating the microscopic field according to Eq. (II.17). A particularly effective way has been provided in the appendix of Ref. 164. Perhaps the most important aspect of the consistent calculations of the macroscopic field is the ability to obtain the local macroscopic dielectric constant. Nevertheless, the ability to obtain the local macroscopic field allows us to analyze experimental studies that involve response to electric fields.

An early example has been provided in the study of Oldfield and co-workers[194], who evaluated experimentally the fluorine NMR spectrum of *Escherichia coli* galactose binding protein and compared the observed shielding to the corresponding calculated results obtained using the calculated electric field. The results are given in Fig. IV.6.

The agreement between the calculated results (which was not accomplished by any earlier calculations) demonstrated our ability to capture the actual local field.

The ability to calculate local fields in proteins can be instrumental in other cases and an excellent example is the study of vibrational Stark effects in enzyme active sites, as discussed in Chapter V.

Although electrostatic energies play a major role in enzyme catalysis, it is a significant oversimplification to simply consider the projection

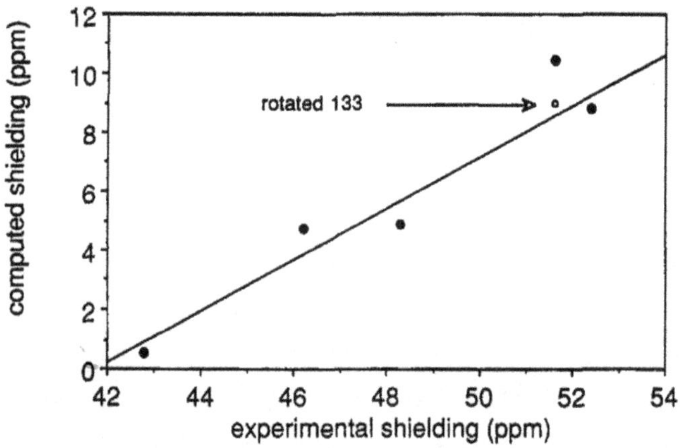

Fig. IV.6 Plot of experimental *versus* theoretical shielding for each [5-F]Trp site in galactose binding protein. The experimental shieldings are those reported in Ref. 4. The computed values are references to a field-free value of 0 ppm. The open circle represents the shielding calculated for a 180° C^{β}–C^{γ} flip of Trp 133. Reproduced with permission[194].

of the electric field on the dipole formed by the change in charges during the reaction. Such a picture may obscure the much more quantitative figure associated with the electrostatic preorganization.

In considering electric fields in biology, it is obviously important to mention the crucial effect of the membrane potential in controlling the directionality of ion transport. The membrane field has been mainly treated by continuum formulation. However, a much more microscopic treatment that includes explicitly the electrodes, electrolyte solution, membrane and ion channels is discussed in Chapter X.

IV.6 How Reliable are the Results of Electrostatic Calculations?

Despite the possible appeal of different electrostatic models, the key issue is whether or not such models are reliable. The issue of reliability (rather than formal appeal) has been appreciated rather slowly, in part because of the use of inappropriate validation approaches. An excellent example is the use of pK_a calculations in validation of electrostatic models. That is, although it has been realized that pK_a value calculations provide an effective way for examinations of the performance of electrostatic models[58,87,93,99,138,167,169,195], most studies have focused mainly on the pK_a values of surface groups. Unfortunately, these pK_a values can be reproduced by many models, including those which are basically incorrect[95]. Thus, the use of equally weighted pK_a values of a large number of proteins as a general benchmark is likely to lead toward unjustified conclusions about the nature of electrostatic effects in proteins. It has been pointed out[95,169] that ionizable groups in protein interiors should provide a much more discriminating picture than the complete set of available pK_a values, because these groups reside in a very heterogeneous environment, where the interplay between polar and nonpolar components is crucial. For example, an ionizable group in a true nonpolar environment will have an enormous pK_a shift, and such a shift cannot be reproduced by models that assume a high dielectric constant in a protein interior. Similarly, a model that describes the protein as a uniform low dielectric medium, without permanent dipoles, will not work in cases where the environment around the ionizable group is polar[52].

The problems with non-discriminating benchmarks were established in Ref. 93 and are illustrated in Fig. IV.7. This figure considers the pK_as of acidic and basic groups and correlates the results obtained by the modified TK (MTK) model with the corresponding observed results. The figure gives the impression that the MTK is an excellent model, but the same agreement would be obtained for any model that uses a large ε since most surface groups have very small ΔpK_a so that Asp and Glu have $pK_a \approx 4$ and all Lys have $pK_a \approx 10$. The same impression has been given by early influential studies of pK_a in proteins (e.g., Ref. 135). Thus, as pointed out in Ref. 93 and emphasized in the inset of Fig. IV.7, we are dealing with a very little amount of relevant information and

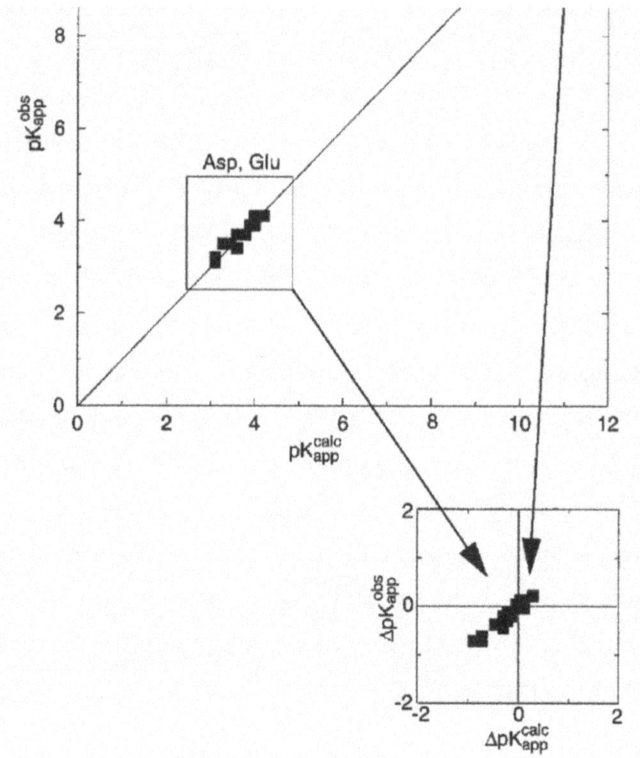

Fig. IV.7 The correlation between calculated and observed pK_a values using a non-discriminative benchmark that only involves surface groups. The figure considers Asp, Glu, and Lys residues on the surfaces of lysozyme and ribonuclease. The calculated pK_a values are evaluated by the MTK(S) model with $\varepsilon_p = 40$ and $\varepsilon_{eff} = 80$. As seen from the figure, we obtain seemingly impressive results for this benchmark, but when the results are displayed as ΔpK_a, it becomes clear that we have simply a poor benchmark where $\Delta pK_a \simeq 0$ and any model that produces small pK_a shifts looks like an excellent model. Reproduced with permission[93].

probably the best correlation would have been obtained with the trivial null model where $\Delta pK_a = 0$ for all surface groups. In view of the above example, it is very important to validate electrostatic models by using discriminative benchmarks with $\Delta pK_a > 2$ and to use the same philosophy for other electrostatic properties. The use of a discriminative benchmark in pK_a calculations[93] has been found to be extremely useful in demonstrating the range of validity and the proper dielectric of different models.

Another example of validation of electrostatic models can be provided by a study of the potential of mean force (PMF) for ion penetration through ion channels[78]. Here, the issue is the validity of seemingly rigorous statistical mechanical approaches such as the PMF of Eq. (III.4). The problem is that such approaches have been examined only by using somewhat irrelevant benchmarks like the torsional potential of alanine dipeptide rather than systems with the proper electrostatic heterogeneity. Furthermore, using actual ion channels cannot be considered a proper benchmark due to the fact that it is not clear what it takes to get converging results in such systems. This problem was addressed in Ref. 78, which used a part of the gramicidin channel with varying numbers of water molecules and established a system that could be used in converging simulations. Furthermore, comparing the PMF for moving through and around the channel (that should be identical) provides a unique way of assessing the hysteresis in PMF calculations.

Chapter V

Electrostatic Origin of Enzyme Catalysis

The search for the origin of the enormous catalytic power of enzymes has been the subject of many studies for more than 100 years. However, the progress in this direction has been very slow since it is very hard to obtain unique analyses of energy contributions using only experimental approaches[196]. Nevertheless, many different options have been proposed as the origin of enzyme catalysis[196]. In this respect it is important to realize that most proposals could not be verified in a unique scientific way. An excellent example has been provided by the attempt to explore the importance of electrostatic effects in catalysis. The relevant experiments were done with model compounds in solution, where it was concluded that electrostatic effects cannot be important[196]. However, this conclusion that was adopted uniformly by the biochemical community appears to reflect the effect of the very large dielectric screening in water and, as will be shown below, it has been a completely incorrect conclusion.

The elucidation of the origin of the catalytic power of enzymes is a subject of major practical and fundamental importance[175,197–199]. The introduction of combined quantum mechanical/molecular mechanics (QM/MM) computational models (e.g., Refs. 197, 199–205) provided a way for quantifying the main factors that allow enzymes to reduce the activation free energies of the corresponding reactions. QM/MM studies, and in particular those conducted by the empirical valence bond (EVB) method[175], provided compelling support to the proposal[206] that electrostatic effects play a major role in stabilizing transition states of enzymatic reactions[196]. For example, all the large catalytic effects illustrated in Fig. V.1 are mainly due to electrostatic transition state stabilization.

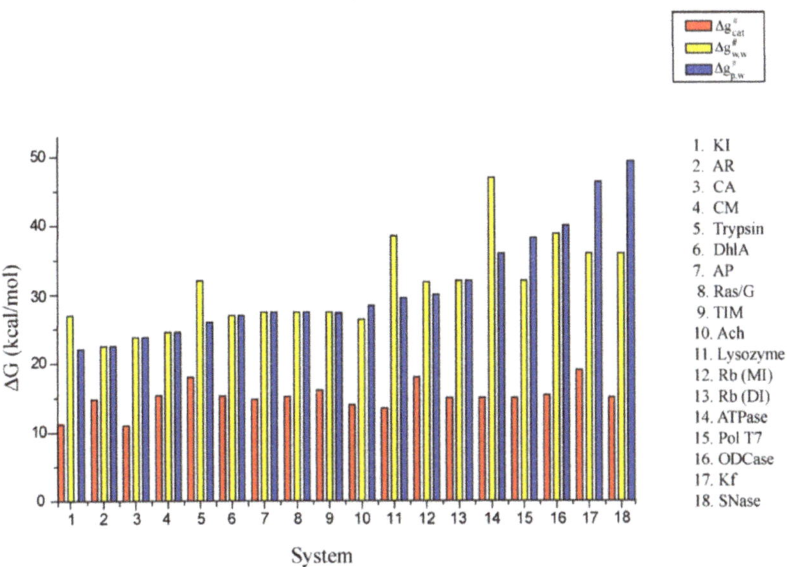

Fig. V.1 Activation free energies of representative enzymatic reactions ($\Delta g^{\ddagger}_{cat}$) and the corresponding reference solution reactions for the same mechanism as in the enzymatic reaction ($\Delta g^{\ddagger}_{p,w}$) and the actual mechanism in water ($\Delta g^{\ddagger}_{w,w}$). The notation of the different enzymatic reactions is defined in Table 1 of Ref. 196. Reproduced with permission[196].

The origin of the electrostatic catalytic effect is somewhat complex as it is due to the change in reorganization energy[196], as illustrated in Fig. V.2.

In examining the above figure it is useful to note that the EVB activation barrier can be approximated quite accurately by the relationship[20]

$$\Delta G^{\ddagger}_{1\rightarrow 2} \approx \frac{(\Delta G_0 + \lambda)^2}{4\lambda} - \left|H_{12}\left(X^{\ddagger}\right)\right| + \frac{H^2_{12}(X_0)}{(\Delta G_0 + \lambda)} \tag{V.1}$$

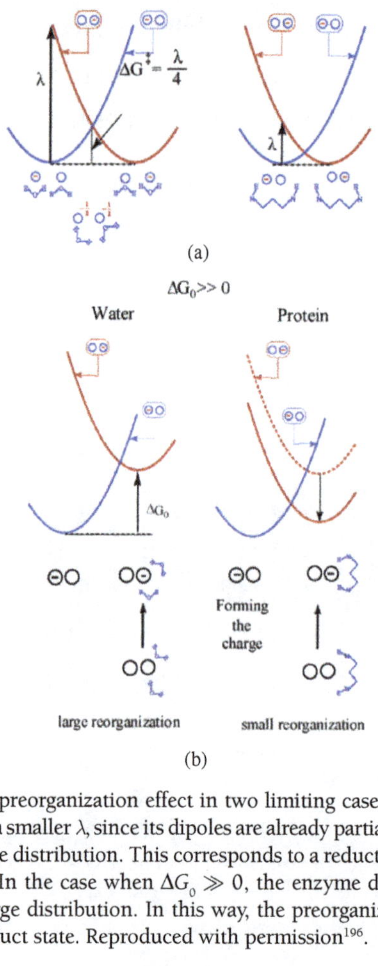

Fig. V.2 Nature of the preorganization effect in two limiting cases. (a) In the limit when $\Delta G_0 = 0$, the enzyme has a smaller λ, since its dipoles are already partially preorganized toward the transition state charge distribution. This corresponds to a reduction of the Marcus "reorganization energy". (b) In the case when $\Delta G_0 \gg 0$, the enzyme dipoles are preorganized toward the product charge distribution. In this way, the preorganization helps to increase the solvation of the product state. Reproduced with permission[196].

The first term is the well-known Marcus activation barrier with the reorganization energy λ, and ΔG_0 is the reaction free energy.

The change in the reorganization energy upon moving from water to the protein active site (that we define as the preorganization effect) is the main reason for the catalytic effect of the enzyme. In fact, the interaction energy of the polar environment and the transition state is similar in the enzyme and in water and the main difference is that in water we have to spend more reorganization energy in bringing the environment polar element to optimal interaction with the transition state.

At any rate, there is now growing appreciation of the importance of the electrostatic preorganization in enzyme catalysis (e.g., Ref. 196 and Refs. 207,208).

Simulation approaches that focus on the electrostatic aspects of enzyme catalysis (i.e., the difference between the stabilization in the enzyme and solution) appear to give much more quantitative results than those that focus on the quantum mechanical aspects of the problem but overlook the proper treatment of long-range effects (see discussion in Ref. 209). Apparently, some problems can be effectively treated even by PB approaches without considering quantum mechanical issues. Some studies focus on the role of the diffusion step of enzyme catalysis[210,211] but (see, e.g., Ref. 212) these effects are trivial compared to the electrostatic effect of enzymes in actual chemical steps. Interestingly, evaluation of activation free energies of enzymatic reactions appeared to be simpler, in terms of the stability of the corresponding results, than other types of electrostatic calculations such as binding free energies (see discussion in Ref. 154). This apparent advantage that has been exploited for a long time in EVB studies (e.g., Ref. 197) has been reflected in molecular orbital QM/MM studies (e.g., Refs. 199, 200, 213).

The view which emerged from all the consistent calculations that actually reproduced the catalytic effect in a quantitative way have found that the major contribution is due in each case to electrostatic effects. These findings are illustrated in Table 2 of Ref. 196.

The idea that enzyme catalysis is largely due to electrostatic effects has been strongly supported by experimental studies. Most notably Boxer and coworkers[214] used measurements of vibrational Stark effect to probe the internal electric field in enzyme active sites. Basically such experiments give information on the change in the product of the local field and the dipole of the reacting system upon going from the ground state to the transition state. This difference gives an estimate of the electrostatic contributions to the change in the activation barrier. The corresponding analysis found a remarkable correlation between the electric field and the catalytic effect. One of the most convincing examples is provided by the study of ketosteroid isomerase (KSI) (e.g., Ref. 214). It illustrates (Fig. V.3) that a major part of the observed catalytic effect is due to electrostatic transition state stabilization. Significantly, the experimental analysis provided a direct support to the careful theoretical studies (e.g., Refs. 215,216) of the catalytic effect of KSI that established the electrostatic origin of this effect. This is important in view of the major, very problematic attempt to use KSI as a proof that enzyme catalysis can be due to countless poorly defined effects but not to electrostatic effects[217].

Influential works have drawn a renewed interest to the effect of distant mutations on enzyme catalysis[218]. Although this was featured as a new paradigm of coupled mode effects[218–220], it is basically a manifestation of the well-known allosteric effect where the change in interactions in a given site changes to some extent the protein structure, and this in turn changes the preorganization in the active site and thus the catalytic effect (see discussion in Ref. 196). Our analysis of allosteric effects[221–224] have identified the transfer of electrostatic effect as the most important factor, and clearly as the best way of correlating the structural changes with the resulting catalytic effect. This point has been formulated[225] in studies of the fidelity of DNA polymerase, where we introduced correlation matrices based on energy coupling rather than on the less informative

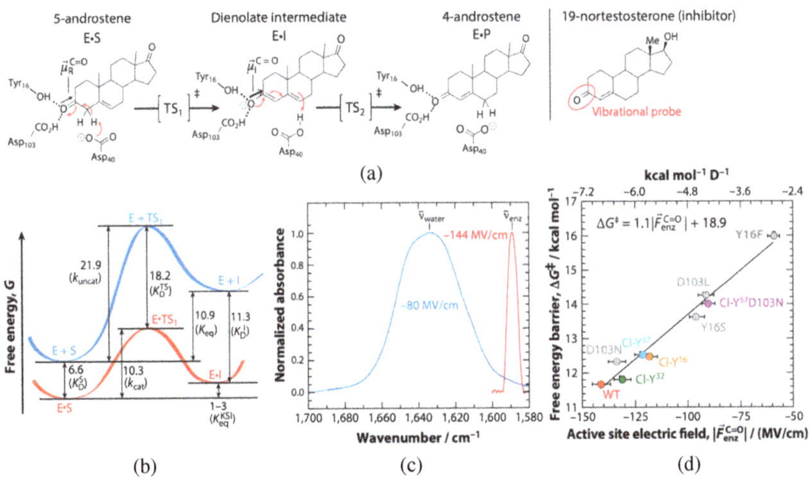

Fig. V.3 Mechanism, energetics, and electric field catalysis of KSI. (a) The steroid substrate 5-androstene-3,17-dione is converted to the conjugated isomer 4-androstene-3,17-dione via enolization (first step) and reketonization (second step) of the carbonyl. (b) Reaction coordinate diagram for the enolization of androstane (S) to the dienolate intermediate (I), either in KSI (red) or in solution (blue). Numbers correspond to free energy differences in kcal/mol. The solution reaction proceeds through a mechanism identical to that of the enzymatic reaction. (c) KSI exerts a very large and homogeneous electric field on 19-nortestosterone's carbonyl bond relative to water's reaction field, based on the significant redshift and band narrowing detected in the infrared spectrum. (d) Linear correlation between KSI's catalytic power (expressed as $\Delta G^{\ddagger}$) and the electric field its active site projects onto the carbon of the inhibitor 19-nortestosterone. Represented are mutations of the oxyanion hole residues to other canonical amino acids (gray points) and subtle mutations to non-canonical amino acids (colored points). Reproduced with permission[214].

structural coupling[218,220]. Our treatment provides useful allosteric matrices that can tell us about the origin of the allosteric energy transfer.

At the end of this section, it is useful to elaborate on my own perspective on the evolution of the importance of electrostatic effects in enzyme catalysis. As mentioned in the beginning of the section, the biochemical community did not believe in the importance of electrostatic effects in enzyme catalysis, and this point can be established by considering the influential writing of Jencks[226], where electrostatic transition state stabilization was not considered as a major catalytic factor. Thus, the electrostatic concept faced significant resistance. Interestingly, in 1989, the renowned enzymologist Jeremy Knowles, who was skeptical of electrostatic catalysis, told me that the longer it took for people to accept the idea, the greater the impact it would have for me — if it were ultimately confirmed — since it would then be harder for others to take credit away from my coworkers and me.

Another interesting story involves a meeting of the American Chemical Society around 2008, where I spoke with Steve Boxer from Stanford, whom I knew well from the field of photosynthesis. At the time, Boxer had begun studying enzymes using vibrational Stark effect measurements and was collaborating with one of my most outspoken competitors, Dan Herschlag. During our conversation, Boxer told me, "*In about a year, you will either love me very much or hate me.*" I replied, "*There is no way you will find anything that consistently contradicts my electrostatic concepts, so I don't think I'll have to hate you.*" As mentioned above, during the next few years, Boxer found more and more detailed experimental proofs that the catalytic effect of KSI (that has been one of the most fierce battlegrounds in the fights between Herschlag and me) is almost completely correlated with electrostatic effects and consistent with our studies. Boxer's findings have landed a major support for the idea that enzyme catalysis is due to electrostatic effects.

In 1978, Max Perutz published what became a very influential review in *Science* under the title "Electrostatic Effects in Proteins". This review[130] featured my works in a major way, emphasizing the idea that enzyme catalysis is due to electrostatic effects. It also implied that the catalysis is due to having charges in a low dielectric, whereas my non-trivial point was that in proteins charges are stabilized in very polar but preorganized environments, but overall it was very supportive. Of course, I wrote Max about this misunderstanding and tried to clarify the problem with ions in low dielectrics.

Gradually, other theoretical workers started to provide support to the electrostatic transition state stabilization idea (e.g., Ref. 227) and to the preorganization concept (e.g., Ref. 228). However, the essence of the preorganization concept remained somewhat obscure to workers without a background in chemical physics. A recent example is the misunderstanding presented in the work of Nome Monger and their coworkers[229]. This work argued (based on difficulties in understanding my Nobel lecture[230]) that my studies ignored the interaction between the enzyme and the reacting substrate, basically reading the statement "*enzyme catalysis is not due to the interaction between enzyme and substrate (which is what was believed by most people), but rather to a large free-energy penalty for the reorganization of the solvent in the reference reaction.*" They completely overlooked the obvious fact that we always calculated the interaction between the enzyme and the reacting substrate but found out that this strong interaction is similar to the corresponding interaction in the reference reaction in solution. Thus the difference in activation barrier appeared not to reflect the small difference in interaction energy but the fact that in solution we have to pay large reorganization energy to reach the transition state, whereas in the enzyme we have to pay much less. Of course, we tried to explain this relatively simple point[231], but apparently it is hard to explain to unmotivated workers.

It is also useful to mention here some incomplete attempts to explore electrostatic contributions to catalysis. That is, some workers tried to calculate the effect of the protein internal field on enzyme catalysis. The first was reported by Johannin and Kellershohn[232], who tried to calculate the effect of the electric field in chymotrypsin. Unfortunately, this pioneering attempt was not so useful, since it did not consider the shielding effect of the water in and around the protein as well as the effect of the protein induced dipoles. It also did not attempt to assess any projection of the field on the change of charges during the reaction. A more sophisticated field calculation was reported by Hayes and Colman[233] but again without considering the shielding by the water in and around the protein.

A related problem appeared in the helix dipole proposal where it was suggested (Section IV.3) and widely accepted by part of the community that the field of the helix macrodipoles has a key role in catalysis. Again, the neglect of the relevant dielectric screening made the estimate very problematic.

The above discussion emphasizes our view that the exploration of electrostatic effects must be based on reliable microscopic energy considerations that include all the relevant factors and not a selected part of the system.

Finally, we would like to point out that even the recent instructive interest in the effect of electric field on catalysis[234,235] is not a substitute for a proper calculation of the actual contribution of the electrostatic free energy on the catalytic effect. This point is demonstrated in the detailed comparative study of the reliability of different approaches for calculating changes in activation barriers[236].

Chapter VI

Redox Potential and Electron Transport Processes in Proteins

Key energy transduction processes in living systems involve electron transport processes. Since these processes involve changes in the charges of the corresponding donor and acceptor, they are controlled by the electrostatic energies of the relevant charges. To understand this control it is important to evaluate the redox and reorganization energies using the relevant protein structure. Probably the first attempt to address this problem was reported in the pioneering work of Kassner[237]. The model used assumed that the protein can be represented as a nonpolar sphere, and this type of model was assumed as a valid model for redox calculations (see discussion in Refs. 76, 238–244). However, the protein dipoles Langevin dipoles (PDLD) studies of Warshel and coworkers have shown[245,246] that redox calculations must take into account the protein permanent dipoles and the penetration of water molecules. The key role of the protein permanent dipoles has been most clearly established in subsequent studies of iron-sulfur proteins[247,248] (see Fig. VI.1).

The effect of ionized groups on redox potentials has been of great interest. It appears that this effect can usually be described well using Coulomb's law with a large effective dielectric constant[99,121,122,188].

Poisson-Boltzmann studies of redox proteins have progressed significantly since the early studies that considered the protein as a nonpolar sphere (see above). These studies (e.g., Refs. 200, 218, 249–252) reflect

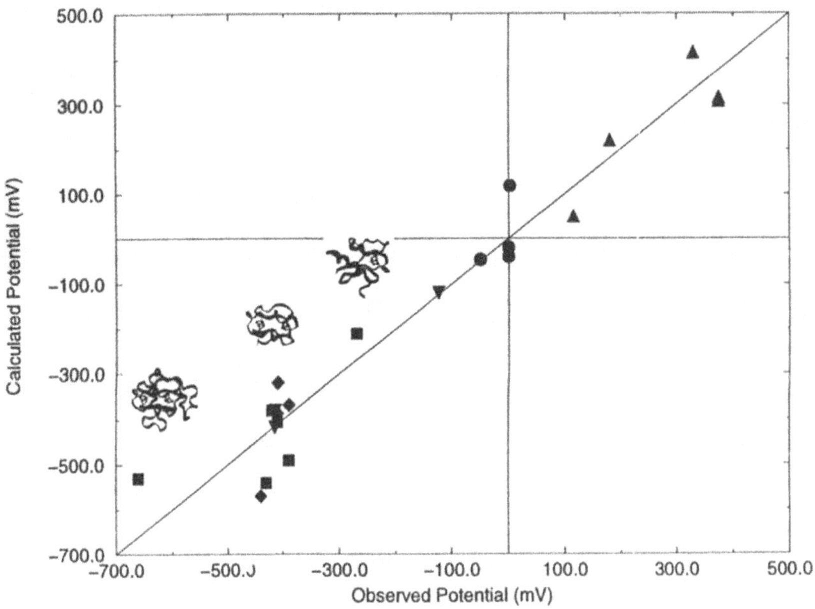

Fig. VI.1 Comparison of calculated redox potentials to the observed. Minimum reduced major axis best fit between calculated and observed redox potentials are shown for: $FeCys_4$ (•); $Fe_2S_2Cys_4$ (◆); $Fe_3S_4Cys_3$ (▼); $Fe_4S_4Cys_4^{1-/2-}$ (▲); $Fe_4S_4Cys_4^{2-/3-}$ (■). The diagonal line shows ideal agreement. Modified with permission[247].

recognition of the importance of the protein permanent dipoles, although some confusion still exists (see discussion in Ref. 76).

Microscopic estimates of protein reorganization energies have been reported[121,183,253,254] and used very effectively in studies of the rate constants of biological electron transport. This also includes studies of the nuclear quantum mechanical effect, associated with fluctuations of the protein polar groups (for a review, see Ref. 40).

One of the most detailed studies of the reorganization of redox proteins upon electron transfer has been provided in Ref. 255. This work used the NMR structures of reduced and oxidized cytochrome c to quantify the reorganization energy and to clarify and define the dielectric constant that corresponds to different operational definitions.

As an excellent example of the overall challenges in studies of the action of redox proteins, it is instructive to consider in some detail computational studies of the primary event in bacterial reaction complexes. The structural elucidation of this system[126,256–259], whose main elements are described in Fig. VI.2, presented at that time a major challenge to both computational and experimental approaches, since it was not clear whether the primary event involved a stepwise electron transfer P^* to B and then to H, or a super-exchange process, where we have a direct transfer from P^* to H^- assisted by the electronic coupling between several states. In the first case, the energy of P^+B^- should be lower than that of P^*B, while for super-exchange P^+B^- should be at a higher energy than P^*B. The careful free energy perturbation/umbrella sampling (FEP/US) microscopic calculations of Ref. 121 (which includes proper boundary conditions and long-range treatment) obtained similar energies for P^*B and P^+B^- and supported the stepwise mechanism. On the other hand,

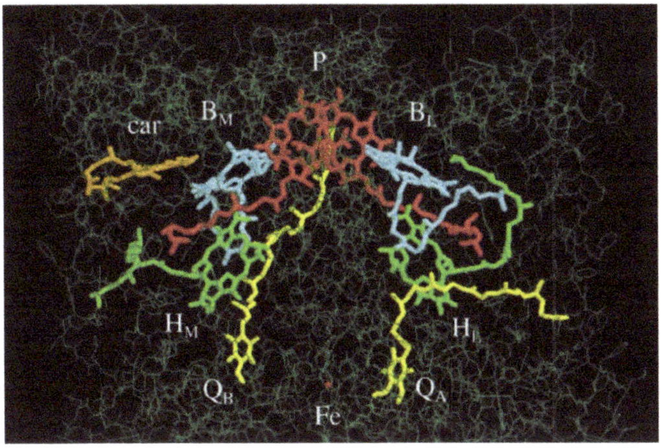

Fig. VI.2 The bacterial reaction center with the key chromophores: the special pair (P), the accessory bacteriochlorophylls (B_L and B_M), the phytochromes (H_L and H_M) and the quinones (Q_A and Q_B). Absorption of light in P initiates an ultrafast electron transfer that lead in about 3 ps to P^+H^-. The puzzle has been whether or not we have a P^+B^- intermediate. Reproduced with permission[40].

inconsistent studies[126,260,261] predicted a super-exchange mechanism. The stepwise mechanism has been confirmed experimentally, but the origin of the difference between the two sets of calculations was not quite clear; subsequent calculations by Marchi, Chandler and their co-workers[126] supported the super-exchange mechanism and argued that Creighton *et al.*[262] did not use a large enough simulation system. Here it has been clarified[121] that the SCAAS model used by Warshel and co-workers provides the results for an infinite system while using a relatively small simulation sphere, whereas less careful calculations that do not include the solvent molecules[126] are incorrectly dependent very strongly on the size of the simulation system.

More recent calculations that used the SCAAS model and longer simulation time basically supported the findings of the previous calculations of Criegton *et al.*[262] and provided further insight on the charge separation process. For example, Fig. VI.3 shows the results of such calculations for electron transfer from P^* to B_A in *Rhodobacter sphaeroides* reaction complexes. The points plotted as "×" and "+" give the results obtained from separate trajectories on the potential surfaces of the reactant (P^*) and product ($P^+ B^-_A$) states, respectively. These free energy curves are most reliable in the regions of their potential minima, where the system spends most of its time during the trajectory. Sampling of the region between the two minima can be improved by combining the results from the two trajectories.

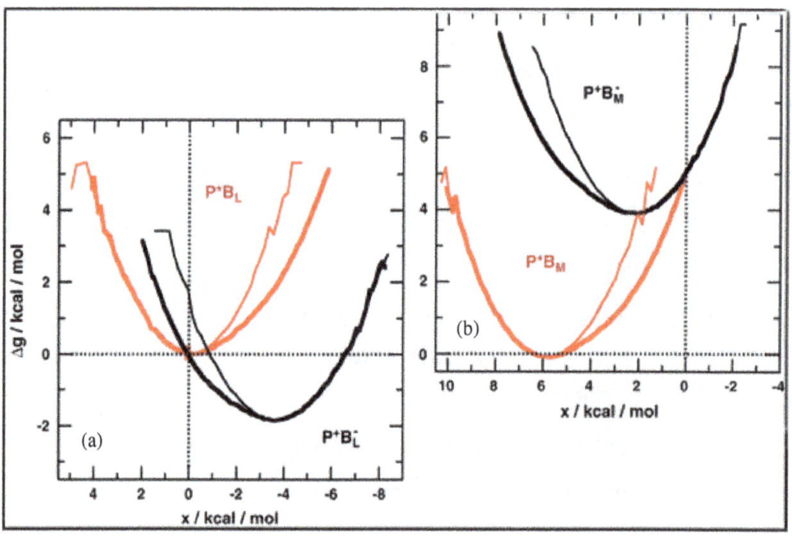

Fig. VI.3 (a) Calculated free energy functions for the diabatic reactant (Δg_{P*}, red) and product (blue) states in the reaction $P{*}B_L \rightarrow P^+ B_L^-$ in *Rhodobacter sphaeroides* reaction centers. The abscissa is the reaction coordinate, which is defined as the electrostatic energy difference between the two states. The free energy functions plotted with thin lines were obtained by the relationship[77] $\Delta g_\beta(x') = -\,k_B T \ln[\langle\delta(x - x')\exp(-x/k_B T)\rangle_\alpha / \langle\delta(x - \langle x'\rangle_\alpha)\rangle_\alpha]$ from separate molecular dynamics trajectories on the two potential surfaces; those plotted with thick lines were obtained by combining the results from these trajectories. (b) Same as (a) but for the reaction $P^*B_M \rightarrow P^+B_M^-$. Adapted with permission[124].

As mentioned above, free energy functionals of VI.3 reproduce the trend of our original study and established that we have a two-step process rather than a super-exchange. It also establishes that the reorganization of the protein is small and that guarantees a fast downhill process that prevents loss of energy by fluorescence.

An instructive issue is the perception that running long simulation time without sufficient focus on long-range problems is superior to early calculations that focused on these issues. A good example is provided by studies of the difference in energetics between the active and inactive branches in reaction complexes. For example, Fig. VI.4 shows the results of similar FEP calculations for electron transfer from P^* to the bacteriochlorophyll on the 'inactive' branch of electron carriers (B_B). The product state (P^+) is found to lie about 4 kcal/mol above P^*. This result agrees with the PDLD calculations on *Rhodopseudomonas viridis* reaction complexes[263] and is in good accord with the observation that electron transfer occurs predominantly to B_A and H_A in preference to B_B and H_B. Mutations that lower the energy of $P^+B_B^-$ relative to $P^+B_A^-$ have been shown to promote electron transfer to B_B and H_B. However, molecular dynamics simulations[250] put P^+ about 7 kcal/mol below P^*. Those calculations differed in using periodic boundary conditions and an Ewald-sum treatment of electrostatic interactions. They also used a larger model of the reaction complex with explicit detergent molecules and water and did not employ FEP, but these differences seem less likely to account for the discrepancy.

Obviously it is very important to examine the validity of the results of consistent microscopic studies of reaction complexes using macroscopic constraints. This is important both for gaining better confidence in

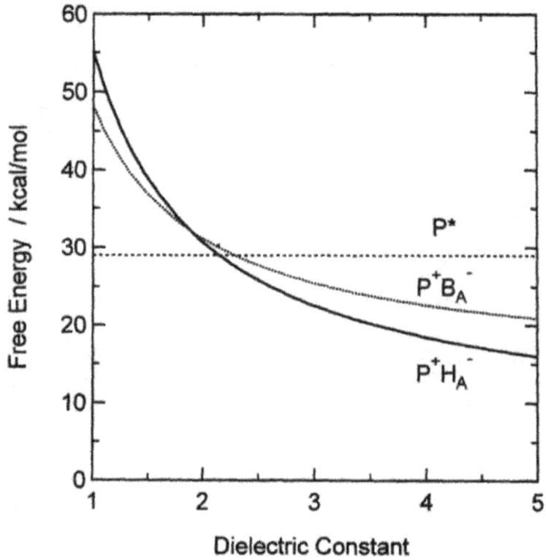

Fig. VI.4 Free energies of $P^+B_A^-$ and $P^+H_A^-$ in *Rhodopseudomonas viridis* reaction complexes as functions of ε in a macroscopic model. All energies in this figure are expressed relative to that of the ground state. The horizontal dashed line shows the energy of the first excited singlet state of P (P*). ΔE_{gas} for the reaction $PH_A \rightarrow P^+H_A^-$ was obtained from the measured potentials of P/P^+ and H_A^-/H_A in purified reaction complexes and PDLD calculations of the change in solvation energy for each of the half-cell reactions *in situ*. ΔE_{gas} for $B_A + H_A^- \rightarrow B_A^- + H_A^-$ was obtained similarly from the reduction potentials of bacteriochlorophyll-*b* and bacteriopheophytin-*b* in polar solvents and calculations of the oxidized and reduced molecules in solution. ΔG_{elec}, the calculated electrostatic free energy changes for $PB_AH_A \rightarrow P^+B_AH_A^-$ and $P^+B_AH_A^- \rightarrow P^+B_A^-H_A$ obtained by Eq. (VI.5), was summed up to obtain that for $PB_AH_A \rightarrow P^+B_A^-H_A$. Ionizable amino acid side chains were taken to be in their neutral ionization states. Reproduced with permission[118].

current theoretical and experimental consensus in the field and in order to see why different macroscopic studies have not pointed out conclusively toward the stepwise mechanism. Thus, we illustrate below the power of consistent phenomenological electrostatic considerations in the case of bacterial reaction complexes.

We start by noting that the free energy change for a reaction such as $PB_A \rightarrow P^+B_A^-$ can be described macroscopically as:

$$\Delta G_{elec} = \Delta E_{gas} + \Delta G_{sol}^{\infty} + \Delta U_{Q\mu} / \varepsilon \tag{VI.1}$$

ΔE_{gas}, as before, is the change in vacuum molecular orbital energies of the electron donor and acceptor, ΔG_{sol}^{∞} is the change in solvation free energies of the electron carriers at infinite dilution in a homogeneous medium with dielectric constant ε, and $\Delta U_{Q\mu}$ is the change in unscreened electrostatic interactions, including the direct interactions between the donor and acceptor (ΔU_{QQ}). As discussed in Chapter II, ΔE_{gas} can be estimated either by quantum calculations or by combining measured E_m values with calculated solvation energies.

We next use Eq. (III.13) to relate ΔG_{sol}^{∞} to the corresponding free energy change in water or another reference solvent with a high dielectric constant, $\Delta G_{sol}^{w,\infty}$:

$$\Delta G_{sol}^{\infty} = \Delta G_{sol}^{w,\infty} + (1 - 1/\varepsilon)/(1 - 1/\varepsilon_w) \approx \Delta G_{sol}^{w,\infty}(1 - 1/\varepsilon) \qquad \text{(VI.2)}$$

This approach is particularly convenient if the same reference solution is used for calculating ΔE_{gas}. Combining Eqs. (IV.2) and (IV.3) with Eq. (II.7), and equating ΔG_{sol}^{ref} with $\Delta G_{sol}^{w,\infty}$, the free energy change for the electron transfer reaction becomes

$$\Delta G_{elec} \approx -F\Delta E_m + \left(\Delta V_{Q\mu} - \Delta G_{sol}^{ref}\right)/\varepsilon \qquad \text{(VI.3)}$$

Calculations using Eq. (VI.3) have been described for the initial ion-pair states in *Rhodopseudomonas viridis* reaction complexes[264]. Fig. VI.3 shows the calculated free energies of $P^+B_A^-$ and $P^+H_A^-$ relative to the ground state as functions of ε. The horizontal dashed line indicates the energy of the excited state (P^*). Experimentally, $P^+H^-_A$ is found to lie about 6 kcal/mol below P^*, or about 23 kcal/mol above the ground state. The macroscopic model returns this result when $\varepsilon \approx 2.9$. (The microscopic PDLD calculations also reproduce the measured energy of $P^+H^-_A$ without using an adjustable dielectric constant but, of course, with much more effort.) More significantly, Fig. VI.3 shows that any value of ε that gives approximately the right free energy for $P^+H^-_A$ puts $P^+B^-_A$ between this state and P*. This is a useful result in spite of the limitations of the macroscopic model, because it provides a back-of-the-envelope check on the PDLD calculations, which also put $P^+B^-_A$ between P* and $P^+H^-_A$.

Perhaps the most important point that emerges from simple models of this type is that the self-energies of charged groups cannot be dismissed as small corrections to the electrostatic free energy. Indeed, they often dominate the problem. In the calculations of the free energy of $P^+H^-_A$ shown in Fig. VI.3, $\Delta G^{w,\infty}_{sol}$ $\left(\Delta G^{ref}_{sol}\right)$ is of the order of –77 kcal/mol, which is similar to ΔE_{gas} and about three times greater than $F\Delta E_m$[257]. Even for the charge-shift reaction $P^+B^-_AH_A \rightarrow P^+B_AH^-_A$ where there is no net change in electrical charge, $\Delta G^{w,\infty}_{sol}$ is about –12 kcal/mol, compared to 4 kcal/mol for ΔE_{gas} and –8 kcal/mol for $F\Delta E_m$[264]. Calculations that neglect the self-energies should not, therefore, be given much credence. The occasional ability of such calculations to reproduce an experimental result can be misleading in this regard, and in most cases probably can be traced to a fortuitous cancellation of errors or to the choice of an indiscriminating benchmark. (See Ref. 52 for a discussion of how calculations by Ref. 126, which omitted the self-energies and gave what is almost certainly much too high an energy for $P^+B^-_A$, fortuitously gave approximately the correct energy for $P^+H^-_A$.)

It might be useful to point out here about the predictive power of proper electrostatic calculations. That is, one of the key questions with regard to the action of bacterial reaction complexes has been associated with the reason for the asymmetry in the system, which allows electron transfer mainly from the A branch while blocking the B branch. Here it has been frequently suggested that the asymmetry is due to differences in electronic overlap between $P^{*}B_{A}$ and $P^{*}B_{B}$ systems. However, our PDLD/S and FEP calculations (e.g., Refs. 263, 264) led to the conclusion that this asymmetry is controlled by energetics, where the polar environment of B_{A} provides much more stabilization to B_{A}^{-} than the less polar environment of B_{B}^{-}. This prediction has been now confirmed more or less conclusively[265,266].

It is also useful to mention recent quantum mechanical evaluation of the redox potentials in blue copper proteins[32], but even in this case it was found that the key problem is the convergence issue, and on many cases semi-macroscopic models give more accurate results than microscopic models.

At this point it may be useful to point out my own perspective on the development of the recognition of the importance of electrostatics in redox systems. As discussed above the first realization of the importance of the electrostatic factor in redox processes in proteins was reported in the pioneering work of Kassner[237], although the model used treated the protein as a nonpolar sphere.

Despite the above advances, the idea that the main control of the redox and reorganization energies are controlled by the overall electrostatic solvation was not widely accepted. For example, in 1982 I attempted to publish in the *Journal of Molecular Biology* a paper that used the X-ray structures of the reduced and oxidized cytochrome c to evaluate the electrostatic free energy along the path between the reduced and oxidized X-ray structures and thus to obtain the corresponding reorganization energy. The paper was rejected (another version was eventually published in Refs. 245, 267) based on the argument that I ignored the major kinetic effect of the protein fluctuations. Of course, the report ignored the fact that the free energy is an average over the fluctuations, and the rate constant is determined by the free energy barrier which is in turn determined by the reorganization energy. Apparently, most scientists at that time did not have a clue about the underlying molecular picture of the microscopic origin of the Marcus theory.

One of the best examples for the above statement occurred in an electron transfer meeting in Philadelphia where I presented a special movie showing the effect of fluctuating dipoles on the electron transfer process. Then I was told by Barry Honig that he went to the restroom after the movie and overheard two leading scientists in the field saying, "What are the stupid dipoles that Warshel is showing?" Obviously, although most people accepted the prediction of Marcus's formula, very few people have any clue on the underlying physical picture.

An additional example of the essential need of proper treatment of electrostatics in studying electron transport systems was provided by a paper[260] of Schulten, Karplus, Michel, Deisenhofer and coworkers. This work adopted our surface hopping approach, but did not figure out how to determine the energy gap between the two surfaces and how to evaluate the electrostatic energy in the protein (e.g., they did not place any water in and around the protein) and thus obtained an enormous error on the calculated energy and predicted a super-exchange mechanism. Of course, this did not reduce the groundbreaking accomplishment of the experimental elucidation of the structure of the reaction complex by Michel, Deisenhofer and Huber, who received the 1988 Nobel Prize in Chemistry, but demonstrated that theoretical studies require focus on relevant quantities. A closely related study was reported by David Chandler and coworkers who also tried to model the primary event in photosynthesis[126], declaring that the conclusions of Parson and myself on this process are wrong. However, as was the case with Schulten and Karplus, they did not include water in and around the protein. They also neglected the "self-energies" the components would have if they were separated to infinite distance in a dielectric medium. Considering the importance of the subject we wrote a detailed response paper[121], pointing out the extremely serious errors in Chandler's work. As expected, he got very upset (despite being the initial offender) and attacked our Langevin dipoles model, which he called "imaginary water". This led us to explain and demonstrate that the Langevin dipoles model represents the physics much more accurately than not including water at all (a treatment that can lead to errors of more than 50 kcal/mol). Chandler then wondered how come those he considered leaders in the field are not using related models and do not subscribe to our dielectric concepts. At that point, we had to clarify that his chosen authorities had very little experience in validating and evaluating the energies of charges in proteins, which happened to be our main direction since 1976. Of course, this did not go well with Chandler, who was a world leader in studying the properties of water, and we were not very successful in communicating to him about the energies of charges in proteins and solutions. The point is that identifying early what the crucial problem is and trying to examine how to treat it is more important that gaining a rigorous knowledge in subjects that are not related to the main problem.

Chapter VII

Electrostatic Contributions to the Energetics of Ligand Binding to Proteins

A reliable evaluation of the free energy of ligand binding can potentially play a major role in designing effective drugs against various diseases (e.g., Ref. 268). Here we have interplay between electrostatic, hydrophobic and steric effects, but accurate estimates of the relevant electrostatic contributions are still crucial.

In principle, it is possible to evaluate binding free energies by performing free energy perturbation calculations for 'mutating' the ligand to 'nothing' in water and in the protein active site. This approach, however, encounters major convergence problems and, at present, the reported results are disappointing except in cases of very small ligands. Alternatively, one may study in simple cases the effect of small 'mutations' of the given ligand[269], for example, replacement of NH_2 by OH (see also Refs. 270, 271).

However, when one is interested in the absolute binding of medium-size ligands, it is essential to use simpler approaches. Arguably the most useful alternative is offered by the linear response approximation (LRA) approach and augmented by estimates of the non-electrostatic effects. That is, the LRA approach is particularly effective in calculating the electrostatic contribution to the binding energy[92,272]. With this approximation one can express the binding energy as

$$\Delta G_{bind} = \frac{1}{2}\left[\left\langle U^{p}_{elec,\ell}\right\rangle_{\ell} + \left\langle U^{p}_{elec,\ell}\right\rangle_{\ell'} - \left\langle U^{w}_{elec,\ell}\right\rangle_{\ell} - \left\langle U^{w}_{elec,\ell}\right\rangle_{\ell'}\right] + \Delta G^{nonelec}_{bind}. \quad \text{(VII.1)}$$

where $U^{p}_{elec,\ell}$ is the electrostatic contribution for the interaction between the ligand and its surrounding; *p* and *w* designate protein and water, respectively; and *l* and *l′* designate the ligand in its actual charge form and the 'nonpolar' ligand where all the residual charges are set to zero, respectively. In this expression the terms $\langle U_{elec,l} - U_{elec,l'}\rangle$, which are required by the LRA treatment, are replaced by $\langle U_{elec,l}\rangle$ since $U_{elec,l'} = 0$. Now, the evaluation of the non-electrostatic contribution $\Delta G^{nonelec}_{bind}$ is still very challenging, since these contributions might not follow the LRA. A useful option, which was used in Refs. 272 and 273, is to evaluate the contribution to the binding free energy from hydrophobic effects, van der Waals, and water penetration. Another powerful option is the so-called linear interaction energy approach.[274] This approach adopts the LRA approximation for the electrostatic contribution but neglects the $\left\langle U_{elec,\ell}\right\rangle_{l'}$ terms. The binding energy is then expressed as:

$$\Delta G_{bind} \approx \alpha\left[\left\langle U^{p}_{elec,\ell\ell}\right\rangle - \left\langle U^{w}_{elec,\ell\ell}\right\rangle\right] + \beta\left[\left\langle U^{p}_{vdW,\ell\ell}\right\rangle - \left\langle U^{w}_{vdW,\ell\ell}\right\rangle\right] \quad \text{(VII.2)}$$

where α is a constant that is around 1/2 in many cases and β is an empirical parameter that scales the van der Waals component of the protein-ligand interaction. A careful analysis of the relationship between the LRA and linear interaction energy approaches and the origin of the α and β parameters is given in Refs. 273, 275.

More implicit models can, of course, estimate the binding energy. This includes the use of the PDLD/S-LRA model for both the electrostatic and non-electrostatic components (see Refs. 272, 273) and the Poisson-Boltzmann approach augmented by estimates of the hydrophobic contributions using the calculated surface area of the ligand.

The idea that conformational averaging is useful for free energy calculations seems quite obvious, generally understood and widely used[199,276]. What is much less understood and somehow still ignored by a large part of the computational community is the realization that a proper LRA treatment requires averaging on both the charged (or polar) and nonpolar states of the ligand. The frequently neglected average on the nonpolar state plays a crucial role in proteins as it reflects the effect of the protein preorganization. The role of this term in binding calculations has been discussed in Ref. 261 and its importance has been illustrated in an impressive way in studies of the fidelity of DNA polymerase[277] (see below).

A powerful combination of the PDLD/S-LRA approach for the electrostatic contribution to binding and the linear interaction energy idea for the non-electrostatic contribution is termed PDLD/S-LRA/β. This approach[275] has been found to be very reliable and effective.

A very popular and widely used method for binding calculations is the semi-macroscopic method, molecular mechanics Poisson–Boltzmann/surface area (MM/PBSA) approach[278].

It predicts the absolute binding free energy by MM energy, solvation free energies with PB or generalized Born (GB) calculations, and entropy estimates from equianharmonic or normal mode analysis. It is worth mentioning that the MM/PBSA and MM/GBSA methods are problematic adaptations of the PDLD/S-LRA idea of molecular dynamics generation of conformations for implicit solvent calculations, because these methods only calculate the average over the configurations generated with the charged solute while ignoring the uncharged term. In the MM/PBSA approach, rather than looking for more physical (and complex) implicit solvent representations, much attention has been devoted to adjusting model parameters to reproduce experimental observations. This led to unjustified compensation of the very large errors in entropy calculations by unrealistic van der Waals contribution. Also, this approach has other problems, such as the treatment of the protein dielectric (see discussion in Ref. 275). Overall, the fundamentally incorrect nature of the MM/GBSA is clearly demonstrated in Ref. 275.

One of the most serious problems in obtaining a converging result in calculations of binding free energies is the fact that the binding process involves a change in the number of internal water molecules and that this process is probably associated with a slow relaxation time. In order to overcome this challenge, we introduced recently the cycle shown in Fig. VII.1, where we actually mutated the ligand to water molecules in both the protein site and in solution. This type of treatment is expected

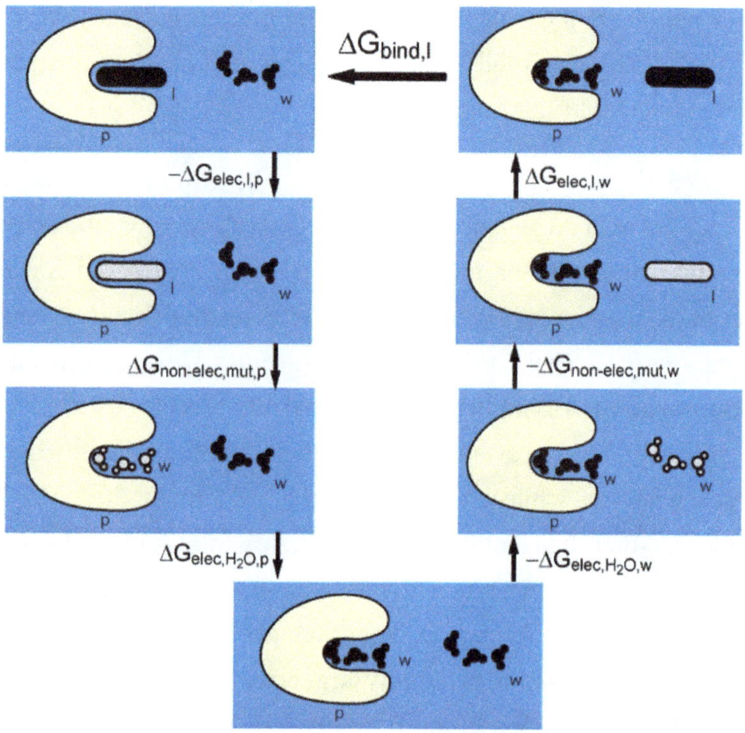

Fig. VII.1 A thermodynamic cycle that deals specifically with the water penetration effect. The cycle involves mutation of the nonpolar form of the substrate to nonpolar water molecules and then to polar water molecules. The polar and nonpolar forms are designated by black and light colors, respectively. Reproduced with permission[124].

to provide a clearer insight on the nature of β and significantly improve the microscopic results.

Another fundamental problem is the evaluation of the binding entropy. Here it was realized that the frequently used harmonic or quasi-harmonic approximations[279,280] are unlikely to give quantitative results, and thus drawing from our early idea[281] and from a related idea of Hermans and Wang[282], we developed a variational restraint release (RR) approach. This approach, which is described in detail elsewhere[283], is more general than Herman's approach, since it is formulated in general Cartesian coordinates, and more importantly, it recognizes the fact that the RR free energy depends on the constraint coordinates and thus minimizes the enthalpic contributions to the RR by searching for the lowest limit of the RR results. This approach has already been used in calculations of binding[284] and catalysis[285,286].

A crucial issue with regard to calculations of binding energies and the energetics of other processes in proteins is the amount of internal water molecules around the given ligand. Here we have to take into account a very delicate equilibrium between the water molecules in and out of the protein and that the corresponding equilibration may require a very long simulation time. One of the most effective ways to deal with this issue is the "water flooding method"[287]. This approach avoids the extremely expensive simulations of evaluation of the energetics of water insertion by a Monte Carlo approach, which can require consideration of an enormous number of rejection of insertion moves. Instead we use a postprocessing treatment to filter out a limited number of configurations from a large possible set. The validity of the water approach has been established by both reproducing water positions in test cases[287] and by systematic implementation of a powerful and efficient grand canonical simulation approach[288].

Chapter VIII

DNA Replication

VIII.1 Background

The reproduction and evolution of life depends on the accurate replication of the genome, which is facilitated by DNA polymerases[289]. The synthesis of new DNA molecules would be impossible without these enzymes as they increase the rate of phosphodiester bond formation by many orders of magnitude compared to the corresponding uncatalyzed reaction in water[290,291]. DNA polymerases are capable of selectively catalyzing the transformation of four structurally distinct substrates. A correct dNTP substrate is selected for insertion by the polymerase depending on the identity of the templating base so that Watson-Crick base pairs are formed with high efficiency. The fidelity of DNA replication is in turn controlled by the polymerase active site where phosphodiester bond cleavage occurs, and by the binding site of the incoming nucleotide paired opposite a template base. The rate of incorporation of an incoming wrong nucleotide is drastically slower than the corresponding rate of the right nucleotide, thereby ensuring high replication fidelity[292–294]. In exploring the function of DNA polymerases, it is essential to address several basic problems, which range from the mechanism of the phosphoryl transfer step to control of the fidelity through modulation of the rate-limiting step. These issues will be addressed below.

The first detailed structure-based information about the mechanism of DNA polymerases emerged from structural studies[295], which discovered the so-called "two metal mechanism". A typical polymerase active site is described in Fig. VIII.1 where the precise reaction pathway and contributions to the catalytic effect are not at all obvious from the crystal

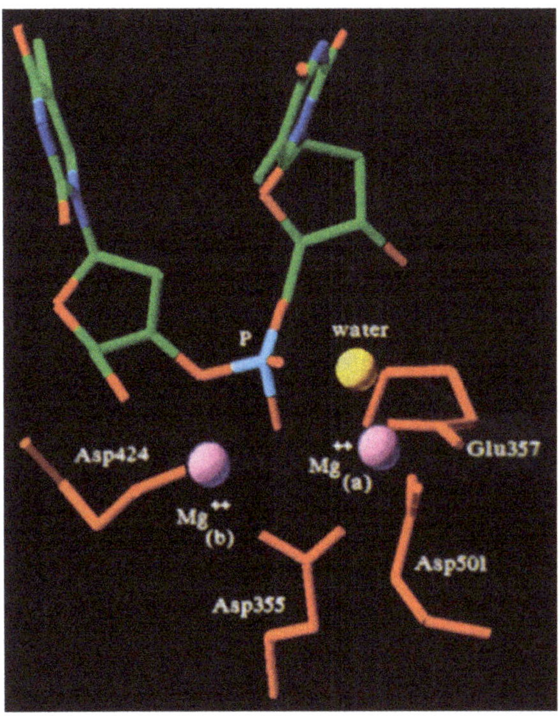

Fig. VIII.1 An example of a typical polymerase active site. Shown here are key residues in the *exo* site of the Klenow fragment, outlining the position of the substrate, nucleophile, catalytic metal ions and key active site residues. Reproduced with permission[113].

structure. That is, the original structural studies were unable to resolve the question of whether the proton on the primer 3′ hydroxyl group goes to the carboxylate of either Asp256 or Glu357 (the authors proposed that there is no need for the participation of an amino acid side chain for nucleophile activation in the chemical step), or what the precise role of the Mg^{2+} cations in the active site is (see also the controversy in Ref. 295). In fact, it was suggested[295] that Mg^{2+} helps the reaction by steric effects, following Westheimer's idea (see discussion in Ref. 113), but the first theoretical study of this system[113] established that the effect of this magnesium cation is entirely electrostatic. Similarly, the theoretical study further clarified problems with the idea that the carboxylate groups of the negatively charged amino acids are stabilizing the attacking hydroxyl group, pointing out that, in fact, the role of these negatively charged residues is most likely simply to position and shield the positive charge on the cationic Mg^{2+}, which will in turn make it much harder for the Mg^{2+} to lower the pK_a of the hydroxide, and therefore that the most likely base is simply just another OH^- anion from the bulk. Subsequent theoretical studies (e.g., Refs. 225, 296–298) have shed more light on the mechanism of DNA polymerases, and this issue will be discussed in greater detail below.

The second key open question has been the control of the rate of nucleotide incorporation by the specific type of base pairing (which, in essence, is the control of fidelity). Here, despite significant experimental milestones in studies of the fidelity of DNA polymerases (e.g., Refs. 293, 299–301), we still do not have a complete quantitative picture of the energetics of this progress. The progress made in using structural information to explore the question of fidelity will be reviewed in Section VIII.5.

VIII.2 The Mechanism of the Nucleotide Transfer

Before embarking on a discussion of the fidelity (and other related issues), it is necessary to analyze the mechanism of the chemical reaction in the protein. In doing so, we assume that the reaction takes place in three generic steps (see Fig. VIII.2). That is, a proton transfer from the 3′ hydroxyl group of the primer is followed by in-line nucleophilic attack of the now anionic oxygen of the primer on the dNTP α-phosphate, with the reaction being completed by the departure of the pyrophosphate leaving group. For the proton transfer step, there are also several possibilities: that is, the proton of the 3′ OH group of the primer could either transfer to the catalytic Asp (e.g., Asp256 in Pol β and Asp654 in T7 polymerase)[296], a non-bridging oxygen of the dNTP α-phosphate (P_α), or a nearby water molecule. Based on previous simulation studies, it would seem that the dominant mechanism involves proton transfer to the catalytic Asp[296,302], or to bulk water, followed by nucleophilic attack on P_α by the automated 3′ OH and the formation of a pentavalent intermediate in an associative mechanism.

Overall, we have demonstrated that electrostatics, rather than orientational effects, drive proton transfer in biological systems, and that proper sampling is essential to obtain the relevant activation free energies. Similarly, we also have shown that when the distance requirements are not extreme since the energy of bringing the donor and acceptor to a close enough distance is not very high, the proton transfer process is highly unlikely to occur via several water molecules. In fact, we are not aware of *any* case where a proposal of proton transfer through several water molecules in catalyzing phosphoryl transfer reaction has

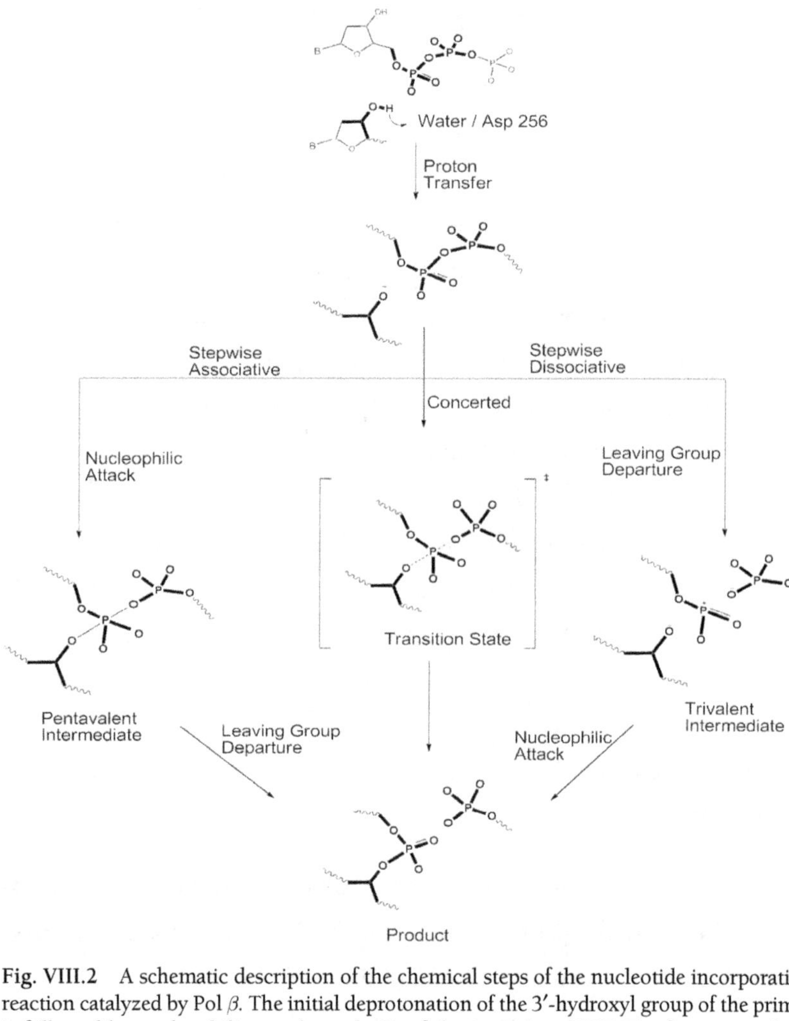

Fig. VIII.2 A schematic description of the chemical steps of the nucleotide incorporation reaction catalyzed by Pol β. The initial deprotonation of the 3′-hydroxyl group of the primer is followed by nucleophilic attack on the P_α of the incoming dNTP. The formation of the $O_{3'}$- P_α bond then results in the release of the pyrophosphate leaving group. As shown in this figure, after the initial proton transfer step, the reaction may proceed via either a stepwise pathway involving a trivalent (metaphosphate) or pentavalent (phosphorane) intermediate, or a concerted pathway which can in turn also be associative or dissociative in nature depending on the extent of bond formation to the incoming nucleophile and bond cleavage to the departing leaving group. However, the hydrolysis of a phosphate diester is unlikely to proceed via a stepwise dissociative mechanism. Reproduced with permission[303].

been shown (by proper sampling) to have a lower barrier than proton transfer through a single water molecule.

VIII.3 Exploring the Fidelity Landscape

Some insight into the fidelity issue has been provided by a study[304] which used computer simulations to model the effect of mutations on the structure of Pol β, and examined the corresponding electrostatic effects by a qualitative analysis. Additionally, theoretical studies of related problems have been reported[305–308]. Molecular dynamics simulations have helped to delineate a possible sequence of conformational changes after dNTP binding in the catalytic cycle of Pol β[309–311]. Simulation studies have also enabled us to test various catalytic proposals for nucleotide transfer reactions[296,312]. Finally, an attempt to progress in a more quantitative direction in the comparison of calculated and observed mutational effects has been reported[302]. This study also introduced a useful approach for constructing interaction matrices and using them to probe the transfer of information between the binding and catalytic sites.

In order to set the stage for our discussion and subsequent analysis, we start with the general working hypothesis of Fig. VIII.3. This conceptual figure has emerged from our studies of DNA polymerases[313,314], in which we concluded (based on calculations of the barrier for the rate-determining chemical step) that the path along the chemical coordinate passes at a somewhat different value of the "orthogonal" coordinate corresponding to the protein structural changes (which is typically considered to correspond to the motion between the open and closed forms). We would like to emphasize that this view is fundamentally different from related assumptions (e.g., Ref. 310) that the barriers for the conformational changes between the open and closed forms contribute to the fidelity or to the reaction rate. That is, the barriers for the motion from the open to the closed form in the reactant state of the R base pair (from $r_{ORS(R)}$ to $r_{CRS(R)}$ in Fig. VIII.3) are unlikely to influence fidelity or, for that matter, the reaction rate for both W and R nucleotide insertion, as long as they are lower than the chemical barrier[315]. Following from this, the interesting possibility that the prechemistry barriers are different for W and R presented in Ref. 310 is not likely to influence fidelity, as long as those barriers are lower than the chemical barrier. In other words, what we are focusing on here is the path from the bound incoming dNTP to the product (i.e., $r_{CRS(R)}$ to $r_{CP(R)}$ and $r_{C'RS(W)}$ to $r_{CP(W)}$), and on

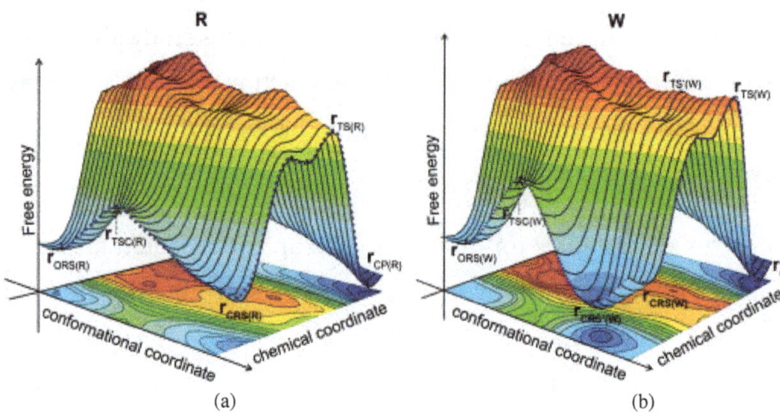

Fig. VIII.3 Schematic free energy surfaces, illustrating potential coupling between the conformational and chemical coordinates, for the insertion of (a) correct and (b) incorrect dNTP by a DNA polymerase. r_O and r_C designate the open and closed configurations, respectively, and TSC designates the transition state associated with the conformational motion. Finally, RS, TS, and PS designate the reactant (enzyme-substrate, ES), transition and product states respectively, and (R) and (W) designate correct and incorrect nucleotide insertion. Reproduced with permission[298].

its dependence on the protein conformation, assuming (based on experimental findings) that the chemical step is rate determining. The relevant precise free energy surface for this can be found in Ref. 298.

VIII.4 Examining the Proposals of Major Prechemistry Barriers

Advances in structural studies have led to several interesting theoretical works (e.g., Refs. 316–318). However, some of these[312,316,317,319,320] have problematic conclusions. For example, a quantum mechanics/molecular mechanics (QM/MM) study of the barrier to prechemistry (before the nucleotide transfer reaction) in human DNA pol β with a G:A mismatch in the active site[317] was estimated to be about 14 kcal/mol. Based on this, the authors suggested that the free energy required for the formation of the prechemistry state is the major contributing factor to the decrease in the rate of incorrect nucleotide incorporation compared with correct nucleotide insertion, and therefore to the fidelity enhancement. Also, in a more recent QM/MM study on DNA polymerase IV[321], it was argued that the prechemical reorganization of the catalytic site (which brings the enzyme from its position in the X-ray structure to an active conformation, prior to the chemical step) would cost approximately 4.0 to 9.0 kcal/mol. However, both these studies employed a treatment that essentially amounts to energy minimization without proper relaxation and sampling. This is problematic for a number of reasons. First, it is important to point out that any QM/MM study has to be carefully calibrated and validated by taking into account the corresponding reaction in water. Secondly, the catalytic landscape for an enzymatic reaction is complex, and without extensive conformational sampling exploring the available conformational space, one usually ends up in a scenario where one gets different (often artificial) activation barriers with different starting structures (for further discussion see, e.g., Ref. 322). In fact, by calculating the potential of mean force for this process using proper relaxation and extensive sampling we obtained a much smaller barrier of 3.58 kcal mol^{-1} to the prechemistry step. More importantly, by using constraints with increasing force constraints from 0.5 to 3.0 kcal mol^{-1} $Å^{-2}$ on all residues within 10 Å of the reactive region, we were also able to reproduce a trend of a correspondingly increasing barrier to this step, highlighting the problems with not allowing the protein to relax properly, and demonstrating that the finding of a high prechemistry barrier is most likely a simulation artifact from the problematic approach used.

In contrast, other workers (e.g., Refs. 294,319) have suggested that intermediates along the path towards the transition state of the incorporation reaction can serves as kinetic "checkpoints". That is, it was basically argued that although the presence of such checkpoints will not change the overall fidelity of the reaction, they *would* define the pathway by which that fidelity is realized. In the case of the polymerase reaction, the intermediate steps (or checkpoints) would allow for the rejection of inappropriately paired dNTPs before the polymerase attempts phosphodiester bond formation of such substrates[294]. Unfortunately, these arguments are not based on clear molecular concepts, and lack support from consistent simulations. In fact, even the common suggestion that the fidelity is due to an induced fit effect is not based on clear structure-energy analysis[314]. However, if one does want to perform such an analysis, one has to first take into account the actual rates of the conformational transition versus the nucleotide insertion step (which stopped flow fluorescence analysis has suggested is ~5× slower than the conformational transition[323]). If the activation barrier for the conformational transition is smaller than that for the chemical step, then it is unlikely to help control the fidelity, and in fact, a detailed study using our renormalization approach[298] demonstrated there is no effect of the motion along the conformational coordinate on the overall rate constant. Here, it is once again important to emphasize that this refers *specifically* to a situation where the chemical step is the rate-limiting step (which is the case with Pol β), and not to a hypothetical scenario where the conformational transition is rate limiting.

Once the observed fidelity has been reproduced, it is then possible to explore the origin of the corresponding difference in barriers. In doing so, we started by examining the idea that this effect involves differences in the so-called "prechemistry" barriers. More specifically, as mentioned above, the study of Ref. 317 estimated the relative barriers in moving the primer terminal O3' atom of the W system from r(ES(W)) to its prechemistry state, using a quantum mechanical cluster model. This study suggested that the prechemistry state corresponds to a local minimum with an energy ~4.4 kcal/mol above the corresponding initial (reactant) state. It was also concluded that the system has to surmount a potential barrier of 14.1 kcal/mol to form the prechemistry state, starting from r(ES(W)). These findings led to the suggestion that the free energy required for the formation of the prechemistry state is a major factor contributing to the decrease in the rate of incorrect nucleotide incorporation as compared to the correct insertion, and therefore to the observed fidelity. Although estimating activation barriers in enzyme catalysis using quantum mechanical cluster models or QM/MM methods and energy minimization approaches is a common practice in the scientific community[312,316,317,320,321,324–326], it has been argued and demonstrated several times in the literature that these models cannot estimate free energy barriers accurately and consistently[322].

Finally, to gain further insight into the catalytic landscape of the R and W systems[298], we generated two-dimensional surfaces for the free energy in the space generated by the chemical coordinate (which describes the nucleotide transfer reaction), and the coordinate describing the movement of the template 3′ primer (see Fig. 9 of Ref. 298). The calculated activation barriers clearly indicate that the nucleotide transfer is more favorable from r(ES(R)), where the primer (nucleophile) strongly interacts with the catalytic Mg^{2+} ion. Furthermore, the chemical barrier from the r(ES(R)) structure is lower for R than for W, reflecting the fact that the R system has better preorganization at r(ES(R)).

VIII.5 Exploring the Molecular Basis for the Fidelity

In order to elucidate the origin of the calculated (and observed) fidelity, we evaluated the binding free energy of the incoming nucleotide at r(ES(R)) and r(ES(W)) in both the R and W systems. The corresponding binding energies are obtained by following the same strategy used in our recent study, where we reproduced the observed change in transition state binding free energy in different mutants and upon changing the base[318]. In this approach, we evaluated the binding energy of the chemical (triphosphate) and the base sites separately by running independent protein dipoles Langevin dipoles simulations using the linear response approximation version (PDLD/S-LRA)[92,327], and using dielectric constants of 40 and 2 for the chemical (triphosphate) and base sites respectively (the justification for this approach is presented in Ref. 318). The corresponding calculated binding free energies reproduced the balance between the binding of the base and the chemical sites, which is the essence of the allosteric effect that controls the fidelity of DNA polymerases. That is, we demonstrated that the fidelity of DNA polymerases reflects allosteric competition between the binding of the base of the incoming dNTP to the base-binding site, and the binding of the transition state to the catalytic site. In the case of incorrect dNTP insertion, poor preorganization in the base-binding site and poor binding of the base lead to structural rearrangements that are propagated to the catalytic site and destroy its preorganization[302]. This allosteric control of fidelity is described in Fig. VIII.4, and this compensating allosteric effect reflects a rather simple interplay of free energy contributions, and should not be described by terms such as "population shifts", since the population shift is simply a result of the change in free energy associated with being in different configurations, and *not* a basis for any well-defined physical effect.

VIII.6 Fidelity is Not Related to Induced Fit

In exploring the role of conformational changes in Pol β, it is important to discuss the idea that the fidelity of DNA polymerase is related to the induced fit proposal[328–332]. The fact that substrate binding induces conformational changes that bring Pol β to its catalytic configuration can be described as an induced fit effect. However, this does not provide any information about the nature of the catalytic effect in the preorganized active site, nor does it tell us how the structure of the ES (or in our case the E-DNA-dNTP) complex determines the activation barrier

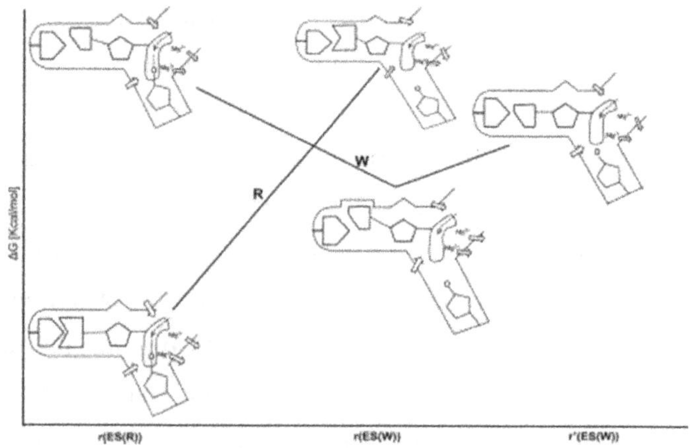

Fig. VIII.4 Origin of the allosteric effect and the relevant changes in the R and W systems at different protein configurations. Specifically, shown here are the protein rearrangement upon binding of R and W at r(ES(R)) and r(ES(W)). In the R case, the protein provides optimal sites for both the chemical and base-pairing parts of the process. On the other hand, in the W case, the protein has to relax in the base-pairing site in order to provide good binding, and this relaxation destroys the preorganization in the chemical part. The diagrams that represent the different configurations illustrate this allosteric effect, with the arrows indicating the protein dipoles. In the R state, these dipoles are preorganized in an optimal way, and thus provide maximum transition state stabilization. In contrast, in the W state, the dipoles are forced to assume less effective preorganization, and thus yield correspondingly less transition state stabilization. Reproduced with permission[298].

of the chemical step. In other words, chemical catalysis does *not* depend on the structure of the enzyme before it binds the substrate. Thus, the induced fit concept does *not* describe the origin of the ES structure. Enzyme catalysis can be considered on the same level as other parts of the folding to ES state (in general the folding energy is invested in reorganizing the active site[333]).

The use of the term "induced fit" to describe the origin of the polymerase fidelity is also quite problematic. In fact, the use of kinetic diagrams of the type considered[331] does *not* provide a unique thermodynamic analysis because they are not based on well-defined free energy surfaces (surfaces of the type presented in Fig. VIII.3). Armed with actual surfaces, we can define the relationship between structural changes upon binding, as well as the corresponding energetics, including the energy change of the enzyme upon moving from r(E) to r(ES) (this energy is defined as the reorganization energy upon binding[255]), the change in the enzyme-substrate interaction upon moving from r(E) to r(ES) and the change in the activation barrier for the chemical step as the result of the binding-induced structural change (this is defined in terms of the change in the reorganization energy of the chemical step and the work term[334]). Without such estimates, it is likely that the discussion will become circular.

As discussed and established[314], it is almost impossible to conduct a logical analysis of the induced fit effect without describing the relevant free energy surface. However, even with a well-defined surface, we must define what the reference state for the induced fit effect is. When this is done[314], we get different results for different definitions. However, none of these results tell us how the fit controls the fidelity.

VIII.7 Ionic Condensation

Nucleic acid molecules are highly charged polyanions. In the folding of nucleic acids, the cations, such as sodium and magnesium ions, are required to neutralize the negative charges on the backbones to reduce the repulsive Coulombic interactions between the phosphates, so that the nucleic acid molecules can fold into compact native structures. Ionic properties such as ion concentration, charge, and size play important roles in determining the stability and folding kinetics of nucleic acids. The effect of ionic condensation around DNA and related molecules has been studied extensively in particular by continuum approaches (e.g., Ref. 335). This is obviously a major electrostatic effect that must be taken into account in modeling structure and function of nucleic acid complexes.

Chapter IX

Electrostatic Contribution to Protein Stability and Folding Energy

Understanding the factors that determine the thermal stability of proteins presents a fundamental and practical challenge (e.g., Ref. 336). One of the outcomes of a better understanding of protein stability should be the ability to predict the trend in stability within related proteins or between different mutants of the same protein. Unfortunately, despite progress in the development of models for studying the folding of proteins[41,337–344], we still have major problems in predicting protein stability by either microscopic or macroscopic models. More specifically, despite the impressive progress in studies of protein folding, we still lack a clear understanding of the contributions of electrostatic effects to thermal stability and to the overall folding free energy. Experimental studies of mesophilic, thermophilic, and hyperthermophilic proteins have provided an excellent benchmark for studies in this area[345]. In general, the number of ionizable residues increases in hyperthermophiles, indicating that charged residues can be considered to be a stabilizing factor. However, some continuum studies have suggested that charged and polar groups lead to destabilization[161,346]. It was also suggested[161] that internal salt bridges tend to destabilize proteins, although as discussed in Ref. 124, this study did not reproduce the relevant observed energies. Other studies (e.g., Refs. 347, 348) appear to support the idea that charged residues can help to optimize protein stability.

The difficulty in reaching clear conclusions on the role of electrostatic interactions in protein stability is associated with the fact that we have a competition between desolvation penalties, stabilization by local protein dipoles, and charge-charge interactions. In the original Tanford-Kirkwood model of a nonpolar protein[349], both isolated ions and ion pairs should become unstable in the "protein" interior[350]. However, the situation is much more complex in real proteins, where charges are stabilized by polar groups[350]. Here the balance between charge-charge interactions and self-energy can depend drastically on the assumed ε_p (e.g., Ref. 124). Our strategy for quantifying the electrostatic contribution to protein stability and to determine its relationship to the assumed dielectric constants appeared to be very effective. We found that we can obtain promising results while using optimal effective dielectric constants for charge-charge interactions (ε_{eff}) and for self-energy (ε_p). More specifically, our approach starts by moving along the (a′) → (c) folding process of Fig. IX.1. In this process we start by folding the uncharged protein and then continue by moving the charges from reference groups in water to the same groups in the protein. Assuming that the folding

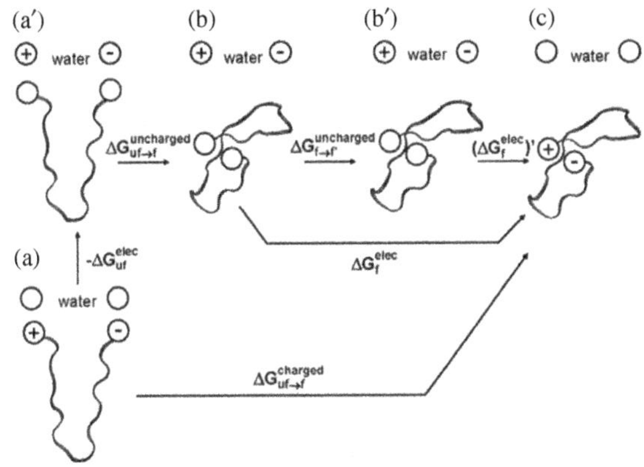

Fig. IX.1 Two alternative descriptions of the folding of a charged protein. The process (a′)→(c) involves a folding of the uncharged protein and subsequent charging, while (a)→(c) involves an initial charging of the unfolded protein and then folding of the charged protein. Structure B is already folded and has the mutant. Reproduced with permission[351].

energy for the uncharged protein is similar for all mutants that involve charged groups, we can focus on the electrostatic contribution to folding. In this case we can use the general expression for the electrostatic energy of different ionization states of a given protein at a given pH[351]. After manipulations that are described in Ref. 351 we obtain

$$\begin{aligned}\Delta G^{elec}_{fold} &= \Delta G^{elec}_{f} - \Delta G^{elec}_{uf} \\ &= -2.3RT\sum_{i} Q^{(f)}_{i}\left(pK^{P}_{i,int}\left(\varepsilon_{p}\right) - pH\right) \\ &\quad +166\sum_{i\neq j}\frac{Q^{(f)}_{i}Q^{(f)}_{j}}{r^{(f)}_{ij}\,\varepsilon_{eff}\left(r^{(f)}_{ij}\right)} + 2.3RT\sum_{i} Q^{(uf)}_{i}\left(pK^{w}_{a,i} - pH\right) \\ &\quad -166\sum_{i\neq j}\frac{Q^{(uf)}_{i}Q^{(uf)}_{j}}{80r^{(uf)}_{ij}}\end{aligned} \tag{IX.1}$$

where *uf* and *f* designate, respectively, the unfolded and folded states. Here, the charge of the ith residue is the intrinsic pK_a ($pK_{a,int}$) of the ith residue in its given protein state when all other residues are neutral,

ε_p is the dielectric constant used in the semi-macroscopic calculation of $pK_{a,int}$, r_{ij} is the distance between residues *i* and *j*, and ε_{eff} is the effective dielectric for charge-charge interactions. The nature of ε_p and ε_{eff} is far from trivial and cannot be fully clarified here. Thus, it is recommended that readers who are unfamiliar with these dielectrics read the discussion in Refs. 93,124. Overall, ε_p determines the intrinsic pK_a and represents the "self-energy" of each charged group. This parameter is not related to the response of the protein to the electric field but to the method used in the calculations and to the elements included in the simulation system. Basically, ε_p reflects all the effects that are not included explicitly in the given model[93]. On the other hand, ε_{eff} is a phenomenological parameter that represents the free energy of charge-charge interactions. This parameter reflects the compensation of the gas phase charge-charge interaction by the reorganization of the solvent and the protein[351]. Of course, the folding free energy includes non-electrostatic contributions such as configurational entropy and hydrophobic contributions and these contributions depend on the path used in Fig. IX.1.

Now we get from Eq. (IX.1):

$$\Delta G_{\text{fold}}^{\text{elec}} \approx -2.3RT\sum_{i} Q_i\left(\left(\mathrm{pK}_{i,\text{int}}^{\mathrm{P}}\left(\varepsilon_p'\right)\right)' - \mathrm{pK}_{i,w}^{\mathrm{w}}\right) + 166\sum_{i\neq j} Q_i Q_j\left[\frac{1}{r_{ij}^{(\mathrm{f})}\varepsilon_{eff}^{(\mathrm{f})}} - \frac{1}{80 r_{ij}^{(\mathrm{uf})}}\right] \quad \text{(IX.2)}$$

It should be clarified that the analysis presented above leads to the conclusion that the folding energetics is strongly modulated by electrostatic effects and that this conclusion is a clear contradiction to

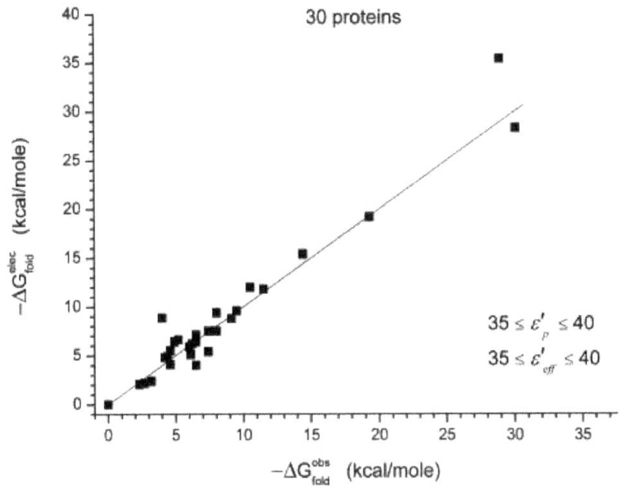

(a)

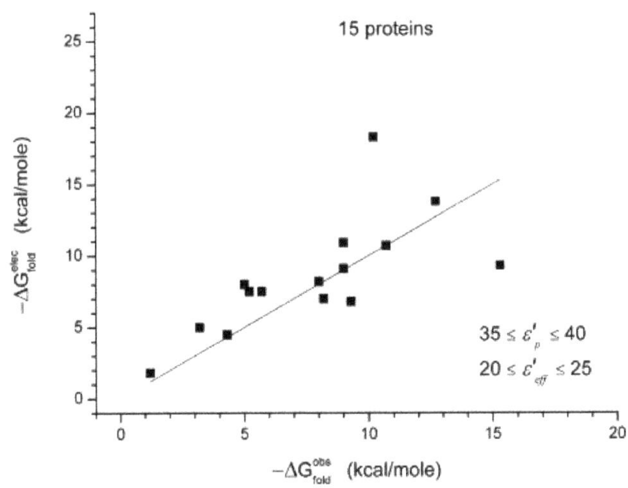

(b)

Fig. IX.2 The correlation between $\Delta G^{\text{elec}}_{\text{fold}}$ and $\Delta G^{\text{obs}}_{\text{fold}}$ after using lower ε'_p for the internal residues. The test proteins fall into two main categories: (a) the one shown in Fig. IX.2(a) where the best fitted $\Delta G^{\text{elec}}_{\text{fold}}\left(\varepsilon'_p, \varepsilon'_{eff}\right)$ is in the range of $35 \le \varepsilon'_p \le 40$, $35 \le \varepsilon'_{eff} \le 40$ and (b) the one shown in Fig. IX.2(b) where the best fitted $\Delta G^{\text{elec}}_{\text{fold}}(\varepsilon'_p, \varepsilon'_{eff})$ is in the range $35 \le \varepsilon'_p \le 40$, $20 \le \varepsilon'_{eff} \le 25$. The majority of the test proteins fall into the first category. Reproduced with permission[352].

previous views (e.g., Ref. 161). Of course, we also have a possible non-electrostatic term but it may be absorbed in the effective dielectric for bringing the charged groups from the unfolded to the folded state in the (a) → (c) path of Fig. IX.1.

The folding energy can be evaluated by the PDLD approach, as was done effectively in Ref. 352, where we optimized the dielectrics ε_p and ε_{eff} to get the best fit between calculated and observed folding free energies. The corresponding results are depicted in Fig. IX.2.

Another effective approach for calculating folding energies has emerged with our coarse-grained model described in Section I.9.

Chapter X

Ion Channels

X.1 Modeling Ion Transport

Many key biological processes are controlled and mediated by the transport of ions through specified transmembrane channels. This includes oxidative phosphorylation, electric signaling in neural and muscular systems (e.g., Ref. 353), etc.

The control of ion transport by ion channels is a complex problem in protein electrostatics (e.g., Ref. 354). That is, we are dealing with ion motion in a free energy landscape that leads from the water solution through the channel, and this landscape is electrostatic in nature as it describes the interaction between a charge and its surrounding and the ion and other ions.

Structural information about biological ion channels was missing for a long time and early studies of ion channels focused on the energetics of ions in the gramicidin channel[355,356] (see also below). Fortunately the landmark solution of the structure of the KcsA potassium channel[357,358] provided a model for real biological channels and an opportunity as well as a major challenge for simulation approaches. Early studies (e.g., Refs. 359, 360) used unrealistic low dielectrics and obtained a major overestimate of the barriers for the K^+ transport. The first careful study was reported by Åqvist and Luzhkov[361] who used free energy perturbation calculations (with local reaction field long-range treatment and the SCSSA boundary conditions) for determining the ion binding at different points along the selectivity filter. Interestingly, early attempts to evaluate the so-called potential of mean force for ion penetration that have the appearance of truly rigorous approaches have not succeeded in reproducing the actual potential of mean force for moving the ions from water to the channel (see discussion in Ref. 134 and below).

Arguably, the first realistic attempt to capture the energetics of the ion penetration through the entire KcsA channel used the semi-macroscopic PDLD/S-LRA approach. This approach coupled with Langevin dynamics simulations was instrumental in providing a reasonable estimate of the activation barriers that actually reproduced the difference in the current between Na^+ and K^+ (see Ref. 171). Interestingly, the selectivity was attributed to a difference in the effective dielectric for K^+–K^+ and Na^+–Na^+. That is, in the case of the smaller ion, the stronger affinity of the protein dipoles to the ions restricts their ability to reorganize and thus leads to a smaller charge-charge dielectric than in the case of the larger ion. Thus, the smaller ions (Na^+) have strong repulsion between them and cannot move with two ions in the selectivity filter at the same time. On the other hand, the larger ions (K^+) move very effectively with two ions in the selectivity filter (Fig. X.1) and thus have a larger ion current.

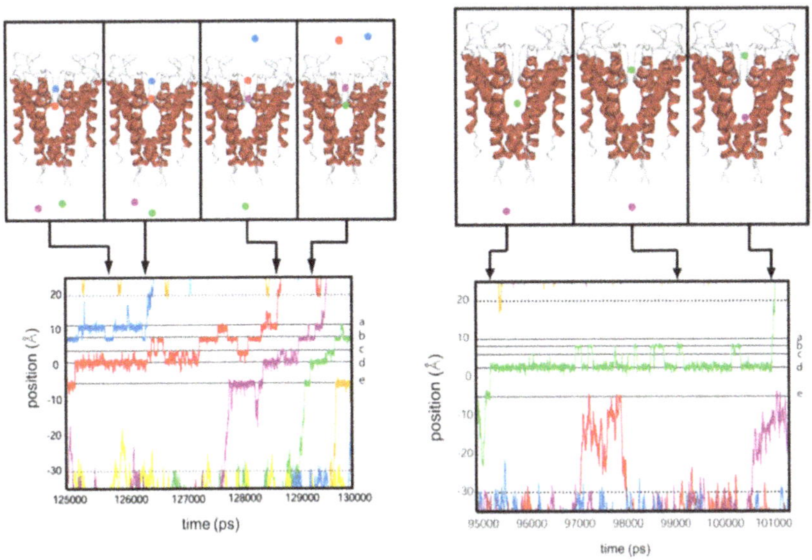

(a) (b)

Fig. X.1 Trajectories of K^+ ions in the KcsA channel. (a) A long trajectory (z as a function of time) that was used in calculating the ion current. (b) The positions of the ions during the trajectory are described by snapshots that depict only those ions which reside in the channel at some period during the time shown. Reproduced with permission[171].

More recent studies with much more computer power started to get reasonable free energy profiles with brute force umbrella sampling (some of the relevant works are mentioned below).

X.2 Membrane Potential and Voltage-Activated Channels

Our need for increasing the physical description of ion channel systems led to the development of a coarse-grained model for electrode/electrolytes, membrane, and membrane/protein systems[362,363]. The model

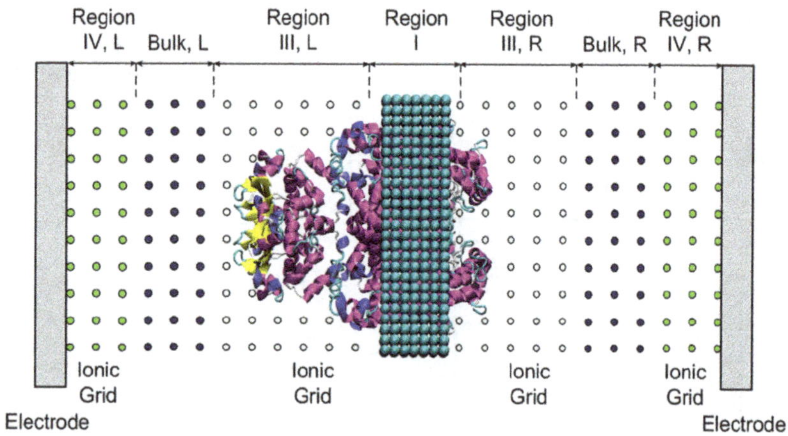

Fig. X.2 A schematic description of the treatment of the simulation system. The regions considered are (i) the protein membrane system (region I); (ii) the region of the solution and the ions that is treated with explicit ions and Langevin dynamics simulation (region II), which is not shown because this region is optional and is not being used in the some studies; (iii) the region with explicit grid points (region III); (iv) the region between the bulk and the electrodes (region IV). The indices L and R represent the left and right sides, respectively. Reproduced with permission[362,363].

(Fig. X.2) represents the key effect of the electrolytes explicitly by placing fractional charges on a grid while solving self-consistently for the interaction between the grid points and the external potential as well as the effect of other grid points. It also introduces consistently and iteratively the effect of bulk regions as well as allows for modeling of the additional periodic cells. The model has been validated by reproducing key test cases[362,363] including the results of a microscopic Monte Carlo study[364] and moved us to a compromise between the continuum models to fully microscopic electrolyte models that involve enormous convergence problems. This allowed us to move to a stage where different factors including the dielectric constants have significantly clearer physical meaning.

We should point out that significant advances have been achieved by other explicit simulation studies (e.g., Refs. 365–370) and by continuum electrostatic models[371–374], but our approach seems to be a very powerful compromise between the convergence difficulties of the all-atom models and the uncertainties associated with the use of continuum models. In fact, the continuum approaches do not include the electrodes and have problems in the formulation of gating current. Thus, although such models are very useful and probably very reasonable, it is hard to use them in clarifying the molecular basis of the action of the ion channel/membrane electrolysis system.

At this point it may be useful to expand on my entrance to the field of modeling ion channels. In the case of ion channels there was some general understanding of the difficulties of moving ions in the channel of a low dielectric[375] and some pioneering works that attempted to model ion channels by a chain of dipoles[327]. However, no attempt was made to capture the relevant energetics in a more realistic model. I became interested in this subject already in 1983, when I visited the lab of Peter Läuger in Konstanz. Peter was a leading physiologist who studied ion channels and died in a tragic climbing accident. I contemplated using my dipolar approach in modeling such channels but did not progress until 1989, when Johan Aqvist, who was my postdoc at that time, used the protein dipoles Langevin dipoles (PDLD) model to explore the free energy profile for ion transfer trough the gramicidin channel[355,356]. The same study also included the first all-atom free energy perturbation calculation of the energetics of ion transfer through ion channels. Although this work had an impact on leading experimentalists in the field (e.g., George Eisenman and Bert Sakmann), it had a limited impact on theoreticians. Most of those who were interested in the field were focused on the possible bumps in the free energy profile that could be captured by subsequent potential of mean force calculations, rather than on the much more challenging and important ability to capture the "solvation" of the ion by the channel; in particular, the ability to capture the change in solvation moving from water to the mixed environment of induced dipoles, permanent dipoles and water molecules inside the channel. Despite intensive collaboration of George Eisenmann with my coworkers, I decided not to spend too much effort on clarifying the problems with other studies that started to be highly quoted, since there was no available structure of a biological ion channel. The situation changed in 1998 when Rod MacKinnon, in a remarkable work, solved the structure of the potassium channel[357,358] (and was awarded the Nobel Prize for that work). On the theoretical front there was a 1999 paper of Roux and MacKinnon[170] who tried to evaluate the energetics of the potassium ion in the channel cavity. This work used a continuum Poisson-Boltzmann (PB) calculation with a far too low dielectric constant that drastically overestimated the role of the helix dipoles. Then in 2000 Johan Aqvist and Victor Luzhkov published in *Nature* a great paper[361] with careful free energy calculations that elucidated key features of the selectivity filter. Subsequent papers of Roux and coworkers used visually impressive but still problematic potential of mean force calculations that work only in cases of a perfect convergence and could not point at problems in the evaluation of the absolute electrostatic energy, the role of ionizable groups and the effect of the protein polarizability. At this stage I decided to focus on deeper understanding of the control of the actual observable, which is the ion current. To do so I turned in 2002[134] to a LD treatment with a channel model that reflected the PDLD electrostatic profile. In a subsequent paper[171] we succeeded in reproducing the ion current of K^+ and Na^+, showing that the K^+ ions move with two ions in the selectivity filter, while the Na^+ ions move one by one (see Fig. X.1). More importantly, we actually elucidated the reason for the selectivity, which could be described as having a smaller effective dielectric for charge-charge interaction for the Na^+-Na^+ pair and a larger dielectric for the K^+-K^+ pair (see discussion above).

In subsequent works we strived to understand the microscopic basis of the voltage-current relationships. This has been a very challenging task, since the available literature focused on macroscopic treatments that have not even considered the electrodes explicitly. Furthermore, the macroscopic treatment did not provide a reasonable and physically convincing description of the gating current (the current that appears at a short time after the change in voltage and before the ions pass through the channel). As in all my previous works I had to move without clear help from the ion channel literature. Thus I felt that I must turn to a microscopic description to be sure of my understanding. In order to progress we built a model which considered explicitly the electrolytes in the solution between the membrane and the electrodes. The electrolytes were represented as a grid of charges that respond to the potential from the electrodes, membrane/protein system and the other grid points. After a major verification effort[362,363] (which involved reproduction of the known analytically solvable test cases) we were ready to explore many key features of membrane-embedded ion channels. This included quantifying the gating current and the meaning of the measured voltage change in response to charge separation in reaction centers (see below). Our ability to explore explicitly voltage effects has become an important tool in other coarse-grained simulations.

At this point it is useful to comment on continuum approaches in studies of voltage current relationships in ion channels. Such approaches have a very long history in establishing the research of ion channels[353]. Ingenious studies (e.g., Ref. 376) provided powerful formulation for the relationship between voltage current and capacity in membrane embedded in electrolyte solution and subjected to external potential. However, the continuum treatment encountered major problems in considering the protein environment with its poorly defined dielectric.

The charge distribution obtained for the membrane system by our microscopic model[362,363] appeared to be different than that obtained in conventional PB studies (e.g., Ref. 376). That is, the PB results are normalized with n_{bulk} rather than by a treatment that guarantees electroneutrality of a finite system, more significantly for the electroneutrality in each side of the membrane. This PB treatment creates a problem in describing the quasi-equilibrium in the short time before the ions pass through the channel where, for example, for a system with an electroneutral protein that was equilibrated before the application of the potential, we should have an electroneutral solution on each side of the membrane (zero net charge on each side) shortly after the potential is turned on. Thus, the formulation used in calculations of the capacitance of the membrane using PB treatments (where the capacitance charges are evaluated by integrating the charge density on the left side of the membrane from $-\infty$ to the membrane surface, e.g. Ref. 376) is formally problematic because the integrated charge is zero in the complete treatment of the system. Apparently the PB treatment does not look at the ions in the bulk or near the electrodes as explicit ions that should be combined with the ions near the membrane when considering the specific charge balance on each side of the membrane. Thus the electrolyte ions that move to the electrodes are considered in the same way that one would consider the electrons in electric wires (namely, these ions are not considered explicitly). We believe that looking explicitly at the electrolyte ions that balance the charge migration toward the membrane do provide an interesting insight.

It is also instructive to note the great insight provided by microscopic Monte Carlo simulations with a primitive model for the electrolytes (e.g., Ref. 377). However, this type of insightful approaches were applied to double-layer model systems, rather than to model systems of a membrane immersed in electrolyte and electrodes, as was done eventually in our study[364].

Finally, we note that the exploration of voltage activation processes led us closer to a microscopic description of complete electric circuits. A notable example is our modeling of voltage change in response to charge separation in membrane proteins[378,379].

Chapter XI

Proton Channels

The spectacular progress in the structural elucidation of proteins that use proton pumps in their energy transduction processes led to advances in understanding key systems such as cytochrome c oxidase[380,381], bacteriorhodopsin[382–386] and ATP synthase[387,388]. Interesting information about the detailed mechanism of proton transport (PTR) has come from careful structural and kinetic studies of the Q_B site of bacterial reaction complexes. In parallel to the progress in structural studies of proton translocation devices, there have been major advances in structural and biochemical studies of ion channels (e.g., Refs. 358, 389–392) and of systems that prevent PTR but allow very efficient water transport (e.g., Ref. 393). Providing a quantitative structure-function correlation for these charge transport systems is a challenge of major fundamental and practical importance. Below we will consider some key examples of PTR simulations.

XI.1 Simulating Proton Transport in Carbonic Anhydrase, Gramicidin and other Systems

Carbonic anhydrase III (CA III) provides excellent benchmarks for studies of PTR in proteins since the step of PTR from the general base (His 64 or its mutants) to the zinc-bound hydroxide (see Fig. XI.1) has been subjected to careful experimental studies. These include the observation of a linear free energy relationship between the relevant ΔpK_a and the rate constant[394].

It was argued that the experiments correspond to the trend obtained by a phenomenological fitting to the regular Marcus formula with a very small and unrealistic reorganization energy[394]. On the other hand, our studies[395] of this system applied the modified Marcus relationship (the first term of Eq. (V.1)) to CA III and were able to reproduce the observed linear free energy relationship without any specially adjusted parameters[395]. We showed that the reorganization energy obtained from the phenomenological fitting is much smaller than the microscopic reorganization energy. This reflected the fact that the regular Marcus

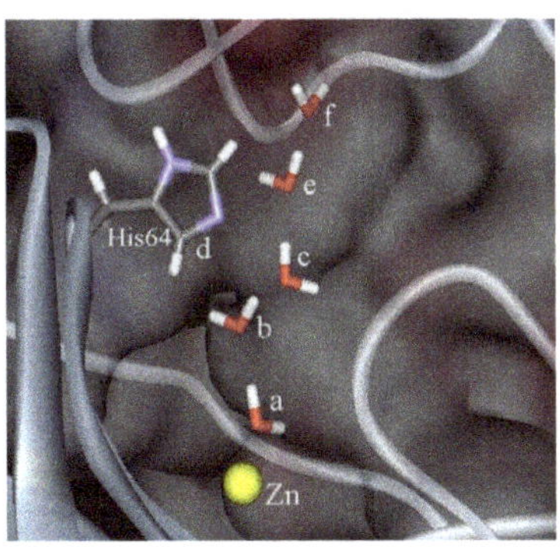

Fig. XI.1 The active space groups in the K64H-F198D mutant of CAIII.

formula does not include the H_{ij} term of Eq. (V.1). Second, it was shown that the process should be described by considering several transfer steps, whereas the phenomenological fitting implicitly assumed a single-step process. The main point that emerged from our study was that the PTR process is controlled by the G_{ij} terms of Eq. (V.1), and thus provides additional support to our modified Marcus treatment. Nevertheless, we subjected this system to more rigorous studies (see below).

Despite the successes of the modified Marcus treatment of CA III, it is important to demonstrate that this treatment is generally valid. This is particularly true in view of the repeated claims that PTR is controlled by orientational effects and the perception that the modified Marcus treatment is only valid for a stepwise process. In order to establish the applicability of Eq. (V.1), we used the more rigorous simplified empirical valence bond (EVB) treatment, where the reacting atoms are described by all-atom EVB while the solvent is treated by Langevin dynamics (LD). In the first step we considered a PTR along a chain of water molecules in water by the full EVB and by the EVB-based LD model. We were able to establish[395,396] that the rate of PTR along a conduction chain depends on the energy of the transferred proton along the chain. This appeared to be the case once we had more than two sites with higher free energy than the previous sites, as was demonstrated by the full EVB model. This finding supports our assertion that PTR in proteins is controlled by electrostatic free energies rather than by orientational effects. In the next step, we used the simplified EVB and LD simulations to explore the PTR in CA III. Our simulations (Fig. XI.2) reproduced the observed rate constants without adjusting parameters for this purpose[396]. This study also gave major support for the use of the modified Marcus treatment since we basically obtained the same results both by using Eq. (V.1) for the stepwise transfer and by using the simplified EVB, which allows for concerted pathways.

Our view on the crucial role of the energetics of the charged states relative to orientational effects has now been adopted by others, including some who originally emphasized more concerted features[397,398].

Although our studies of aquaporin (see below) and CA III indicated that PTR in proteins is controlled by the electrostatic barrier, it is important to examine this finding in systems with a relatively small barrier where orientational effects can be important. (Note that the reorganization of the environment is considered explicitly in our treatments.) In the limit of zero ΔG_{ij} (e.g., PTR in water), the rate obviously is limited by the Grotthuss mechanism and the corresponding orientational effect, but at some point when ΔG_{ij} starts to be significant, the rate begins to be controlled by the electrostatic barrier.

The same strategy was used in a very systematic study of the PTR in Gramicidin[399] (which is arguably the most reliable test case). Remarkably, the simulations reproduced the observed PTR rate and proton current. Our study indicated clearly that also in this case the electrostatic barrier controls the PTR process.

Our studies of PTR in proteins also considered the PTR network in bacterial reaction centers[400] using the PDLD/S proton energy and then using a master equation for propagating the protons.

Another system where we performed systematic studies of PTR is cytochrome c oxidase[45] (see also Section XV.1). This study was based on the time-dependent Monte Carlo approach of Eq. (I.25). Other studies (e.g., Ref. 401) focused on the energetics and pathways of the PTR.

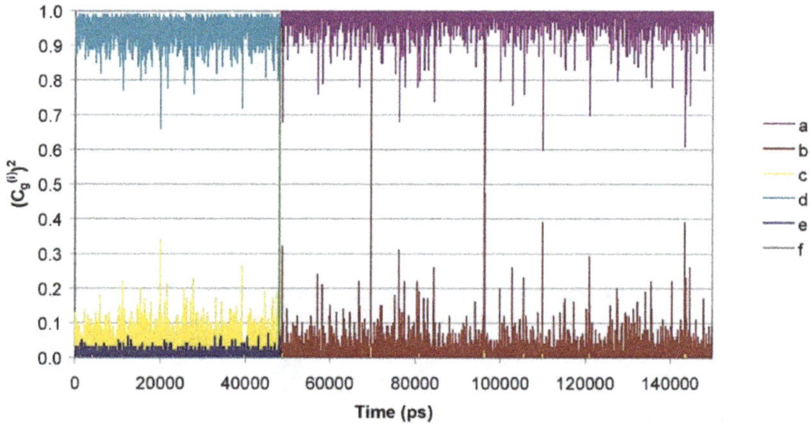

Fig. XI.2 The time dependence of the probability amplitude of the transferred proton for an LD trajectory for a PTR that starts at His 64 and ends at OH^- in the K64H-F198D mutant of CA III. The chance of being on each site is designated by the color code given on the right end. The calculations were accelerated by considering a case where the minimum at site d is raised by 1.2 kcal/mol. Reproduced with permission[396].

XI.2 Aquaporin

Aquaporins play a remarkable role in controlling the conduction of water molecules through cell membranes (e.g., Refs. 393, 402, 403). The study of aquaporins led to the puzzling finding that these channels block PTR, and the origin of this effect has been a subject of major interest[172,173,404–412] and significant controversy. The controversy reflected some misunderstanding and unfamiliarity with the progress in electrostatic calculations. Most early works attributed the barrier to special structural elements[413,414] and, in particular, to the so-called NPA motif[404,407,412], to ionized residues[412], and/or to the helix dipoles[172,405]. All of these factors were supposed to interfere with the water orientational effects that disrupt the Grotthuss mechanism. On the other hand, Burykin and Warshel[410] (referred to hereafter as BW) concluded that although the electrostatic barrier reflects all the contributions of the channel (polar and nonpolar groups), the barrier remains very high even when these contributions are removed, reflecting the relatively nonpolar interior of the channel.

A very problematic analysis which reflected a major misunderstanding of the BW proposal as well as of key electrostatic concepts was presented by Grubmüller and coworkers[415] (after finally realizing the importance of the electrostatic barrier). This work presumed to describe the different proposals for the proton blockage by aquaporin similar to Fig. XI.3, but did so in a very misleading way, which is summarized and clarified in the captions of Fig XI.3.

The captions of Fig. XI.3 clarify the problems with the convoluted analysis of Grubmüller and coworkers, but more importantly the origin of the barrier for PTR in aquaporin was examined by careful systematic works[174], using semi-macroscopic and microscopic calculations. We explored the effect of different factors emphasizing the reliability of the calculations and the problems with alternative approaches.

The calculations summarized in Fig. XI.4 have demonstrated that the effect of the NPA motif is much smaller than what would be concluded from semi-macroscopic calculations that do not use the LRA approach and hold the protein fixed during the mutation of the NPA residues to their nonpolar form. The same types of results were obtained for the mutation of the helix dipoles. Other calculations illustrated the problems with the microscopic potential of mean force calculations that attempted to determine the effect of the mutations by pointing out that these calculations did not evaluate the work of moving from the bulk to the channel interior.

Our work has established that the barrier for PTR in aquaporin and in proteins in general is determined by the desolvation barrier which can

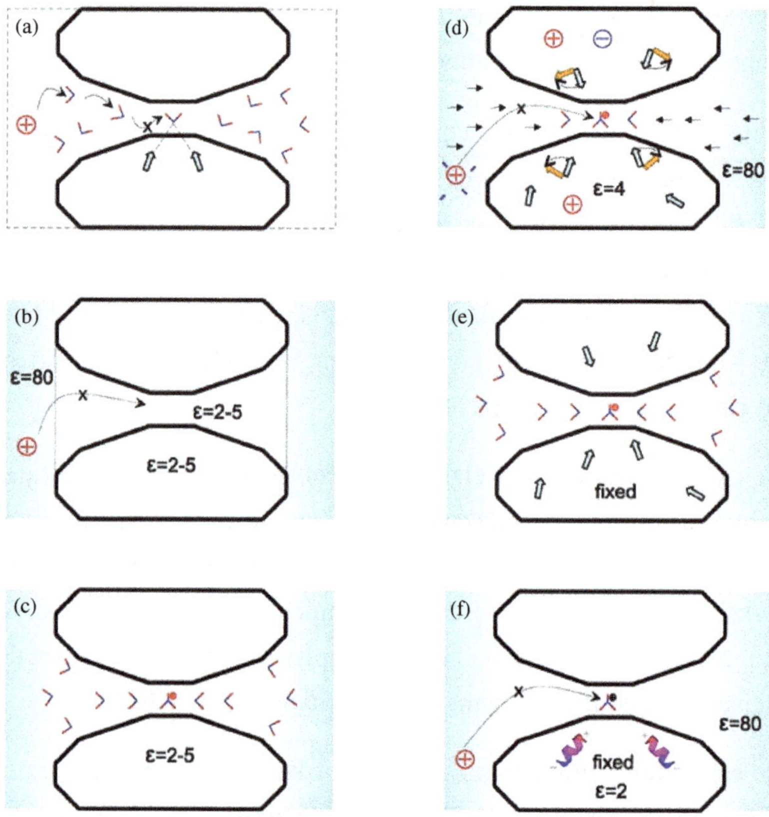

Fig. XI.3 A schematic description of different proposals for the origin of the barrier for PTR in aquaporin. The figure presents: (a) the well-understood view that the barrier is due to the orientation constraint imposed on the water molecules by the NPA motif[405,409]; (b) the misunderstanding of the BW model[410,411], where the channel is presumably described by a low dielectric environment[196] (a view that was never proposed by BW, who actually emphasized that the overall effect of the channel can be represented by a relatively high effective dielectric of $\varepsilon_p = 7$, when the dielectric is described by the effective solvation); (c) the same model as in (b) above but now with the more reasonable assumption that the pore is represented by a high dielectric or by a simplified solvent model, thus presenting a slightly less incorrect view of the BW model; (d) the actual proposal of BW, who considered the complete set of all the electrostatic elements in the channel in a consistent way and examined the overall effect in substituting the solvation by bulk water; (e) the same as in model (d) but with fixed protein (i.e., the model does not allow the protein dipoles to rearrange during the charging process (e.g., Refs. 173, 408)); and (f) the idea that the barrier is due to helix dipoles[196,416]. Reproduced with permission[171].

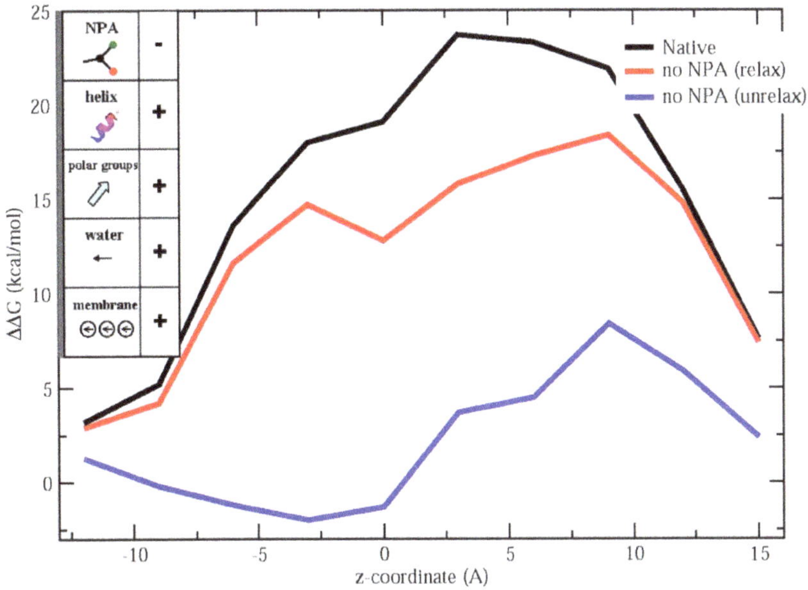

Fig. XI.4 Effect of the NPA residues on the electrostatic barrier by the PDLD/S-LRA approach. The figure provides both the results for mutations of the NPA residues and the inconsistent results obtained by PDLD/S calculations that average only the first term in Eq. (III.2), when the protein does not relax during the mutation process. The fact that the NPA region is mutated is designated by the minus sign in the corresponding box. Reproduced with permission[174].

be compensated for by specific electrostatic interactions. The barrier can be flattened only when the sum of the electrostatic contributions from the protein permanent dipoles, the induced dipoles and the charges are as large as the solvation in water.

The overall perspective on the development of the field might be clarified by elaborating on my personal involvement in proton transfer calculations. In 1978 I started to be interested in proton transfer in bacteriorhodopsin. This was in part because of my longstanding interest in rhodopsins and in part because of a request to review a paper of John Nagel (eventually published as Ref. 126) on the pumping of bacteriorhodopsins, where I argued that the paper led to a perpetual motion machine since it ignored the rate constant of the back reaction (where the proton goes back from the membrane to the protein). My problems with Nagel continued since as a physicist he was not aware of linear free energy correlation between the proton transfer activation barrier and the transfer free energy. Instead, he promoted the idea of the Grothuss mechanism (see above and Refs. 417, 418) that was accepted by most of the community and even featured later in the Nobel Prize talk of Peter Agre, who said subsequently that he got the prize not for understanding proton transfer but for the discovery of aquaporin.

At any rate, at the time I was probably the only scientist who calculated proton transfer in solution or proteins and thus had a very different perspective on this issue than a large part of the scientific community, although physical organic chemists were very aware of the linear free energy relationship between pK_a differences and the rate of proton transfer processes. Overall, my works on PTRs were largely ignored until around 2004, when most of the scientific community was obsessed with the Grotthuss mechanism (where the proton motion is almost barrierless), ignoring the solvation barrier of moving the proton in different sites of the protein[174].

As in the case of the bacterial reaction center, the problem was that most workers in the field did not have previous experience with the evaluation of solvation energies in solutions and proteins, nor any experience in proton transfer calculations. Thus they were ready to adopt the idea of the Grotthuss mechanism that is relevant to proton transfer in solution (where the energy of the protonated water at each site is identical) but not in proteins. In the case of aquaporin almost the entire community was oblivious to the solvation barrier until our BW paper and then the story of many workers suddenly changed.

XI.3 Voltage-Activated Proton Transport

One of the exciting examples of electrostatic control is the action of voltage-activated proton channels (e.g., Refs. 419–428). The understanding of such systems has important medical implications (e.g., Ref. 420) and it also presents a major simulation challenge (e.g., Refs. 429–431). A significant progress in this direction has been provided by studies of the human voltage-gated proton channel, Hv1. Here it is particularly interesting to consider simulations of the voltage activation process, based on the relevant structural model (Fig. XI.5).

Combining our experience with modeling PTR and voltage activation channels[378,433–435] we performed what is arguably the most relevant study[432] that addresses all the relevant issues (PTR, voltage activation, quantum mechanical treatment of a tight proton transfer, and more). Such a comprehensive approach is crucial since we cannot obtain any real understanding by just exploring a part of the problem, as was done

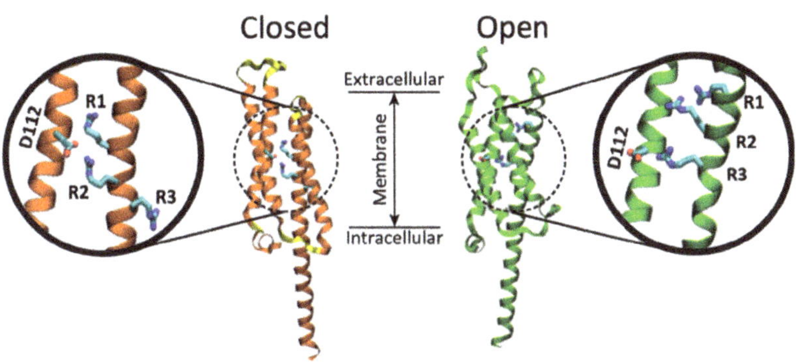

Fig. XI.5 Chimera models used in the study. D112 and R1-3 are shown in licorice representation. The membrane span is shown in black lines. In the closed model, the parts shown in orange are from the original X-ray whereas the parts shown in yellow are the missing loops added by MODELLER. Note the movement of the Arg residues with respect to D112 upon activation in the zoomed-in circles (some helices were omitted in the zoom-in for clarity). Reproduced with permission[432].

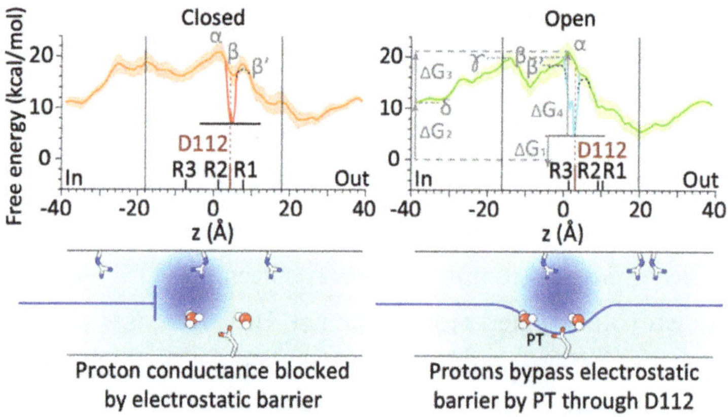

Fig. XI.6 The PDLD/S-LRA energy of the hydronium ion is shown as a function of the position throughout the channel in closed and open conformations of the chimera models. The positions of the ionizable groups of D112 and the gating Arg residues are shown as short vertical lines. The membrane boundaries are shown as long vertical lines, and the energy of the protonated D112 state is shown as a horizontal line. The supplementary lines around D112 (magenta and cyan) represent the reduced-coordinate PTR energy. However, the EVB results for the PTR energy are adjusted according to the corresponding PDLD/S-LRA results. An illustration of the gating and selectivity mechanism is shown at the bottom, where the blue circle represents the electrostatic barrier and the blue line represents the conductance of protons. Reproduced with permission[432].

in simulation studies (e.g., Refs. 419, 428, 436, 437) that did not model the actual PTR energetics and the effect of the voltage activation.

Our study provided a detailed description of the exquisite use of an ionizable aspartic acid residue and arginine residues in controlling the crucial selectivity of the PTR through the channel (Fig. XI.6). Capturing this effect requires the ability to simulate the overall electrostatic potential across the channel and the quantum mechanical proton transfer energetics of passing through a proton acceptor.

Chapter XII

Protein-Protein Interactions

Protein-protein interactions play a crucial role in many biological systems, including in signal transduction[438], energy transduction[439], different assembly processes (e.g., Ref. 68) and electron transport[440,441]. Here again, it seems that electrostatic effects are the most important factor in determining the nature and strength of the interaction between protein surfaces.

Calculations of protein-protein interactions are very challenging, particularly because of the large surfaces involved and the large structural changes upon binding. Considering the protein reorganization upon binding is particularly challenging and a linear response approximation study that tried to accomplish this task for the Ras/Raf system[177] did not succeed to obtain quantitative results. However, a much simpler approach, which evaluated charge-charge interaction by the effective dielectric constant of Eq. (III.10), gave encouraging results for the effect of mutations of the binding energy between the two proteins[177].

Studies of Barnase and Barstar[161,393] also indicate that the results are very sensitive to the dielectric model used. This study identified problems associated with the use of Poisson-Boltzmann with different standard dielectric constant models.

An instructive study of protein-protein interaction will be described in Section XIII.4 which considers the interaction between Ras and Raf. This case is not only very instructive but also offers a way to innovative therapeutic approaches. Coarse-grained modeling offers a very effective way of assessing protein-protein interactions. Examples are provided in Chapter XV where we consider G-protein coupled receptors. Further examples are provided in studies of the interanion between the spike proteins, as well as evaluation of the most important mutations that determine the emergence of new variants. These studies also include evaluation of the interaction between antibodies and different variants[442]. This work used a coarse-grained model (that was used previously in predicting the importance of mutations of N501) to calculate the free energy change of various types of single-site or combined-site mutations. This was done for the UK, South Africa, and Delta variants. We investigated the underlying mechanisms of the binding affinity changes for mutations at different spike protein domains of SARS-CoV-2 and provided the energy basis for the resistance of the E484 mutant to the antibody m396. Other potential mutation sites were also predicted. Furthermore, the *in silico* predictions were assessed by functional experiments. The results establish that the faster spreading of recently observed mutants is strongly correlated with the binding-affinity enhancement between virus and human receptor as well as with the reduction of the binding to the m396 antibody. Significantly, the current approach offers a way to predict new variants and to assess the effectiveness of different antibodies toward such variants.

Chapter XIII

Signal Transduction

The action of the cell is controlled and orchestrated by the transduction of signals (Fig. XIII.1). Representative systems involved in signaling pathways include G-proteins, kinases, cyclic adenosine monophosphate and several phosphatases. Due to the breadth of the field, the present section will consider only a few typical examples, and discuss the factors that control the signal transfer.

XIII.1 G-Proteins

Nucleoside triphosphates serve as crucial mediators of life. For instance, the regulation and control of signal transduction and transport processes relies almost exclusively on GTP. ATP, in turn, is used to drive unfavorable chemical reactions, to fuel biological machines, and to regulate key biological processes (via the phosphorylation of proteins). Other nucleotides also play important biological roles. Their action is controlled by proteins that tune their binding affinity, and the rate of their relevant chemical transformation (see, for example, Refs. 443–447).

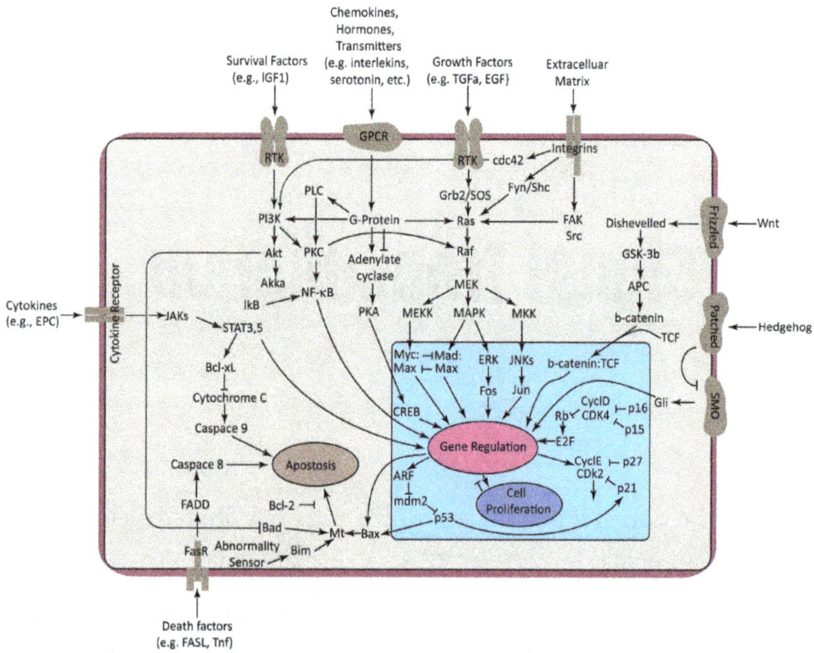

Fig. XIII.1 An overview of cellular signal transduction pathways, highlighting the central role of Ras. Adapted from the Wikimedia Commons file "Image: Signal_transduction_ pathways.png", public domain.

G-proteins are a superfamily of regulatory GTP hydrolases that form a relatively stable complex with their substrate (GTP) and product (GDP). The binding to and hydrolysis of GTP by these proteins triggers conformational changes, which are used in order to communicate with other proteins. The GTP and GDP bound conformations generate the active (ON) and inactive (OFF) states of the protein, where the transformation between these states allows the system to serve as a regulatory machine. The transition between the ON and OFF configurations of these systems can be controlled in a variety of ways, and the G-protein superfamily includes switches (Ras and its homologues), clocks (heterotrimeric G-proteins and subunits) and sensors (elongation factors such as EF-Tu and EF-G). The exchange of GDP and GTP is also regulated by special proteins (guanine exchange factors — GEF).

Despite great advances in studies of G-proteins and related systems (e.g., Refs. 443–447), we still do not have a complete understanding of the detailed molecular mechanism of these proteins. For example, the reasons for the oncogenic effect of some Ras mutations are not clear. Further progress in this field requires quantitative structure-function correlation, which would allow for the conversion of the available structural information into activation energies and, therefore, into the relevant catalytic effects.

One of the best examples of this class of proteins is p21ras. This protein plays a central role in the signal transduction pathway that controls cell proliferation, and can be considered as a switch, where the signal is ON in the GTP-bound form, and OFF in the GDP-bound form[448–450]. As illustrated in Fig. XIII.2, the switch is activated by the binding of GTPase-activating proteins (GAPs), which accelerate the relatively slow GTP hydrolysis in Ras by up to five orders of magnitude. Mutations at either the 12, 13 or 61 positions drastically slow down the GTPase reaction of the Ras-RasGAP (RasGAP) complex (Fig. XIII.3), leaving the signal in the ON state, and thus frequently leading to cancer. The elucidation of the structure of Ras and the RasGAP complex[451] has provided what is perhaps the most detailed link between protein structure and oncogenic mutations. A spectacular clue about the control of the Ras switch was the finding that Arg789 of GAP, which is also referred to as the "arginine finger", provides major transition state stabilization during the GTPase reaction[451]. However, despite this finding (and other major advances), it is still unclear what the origin of the anticatalytic effect of the oncogenic mutations of Ras actually is, with the most

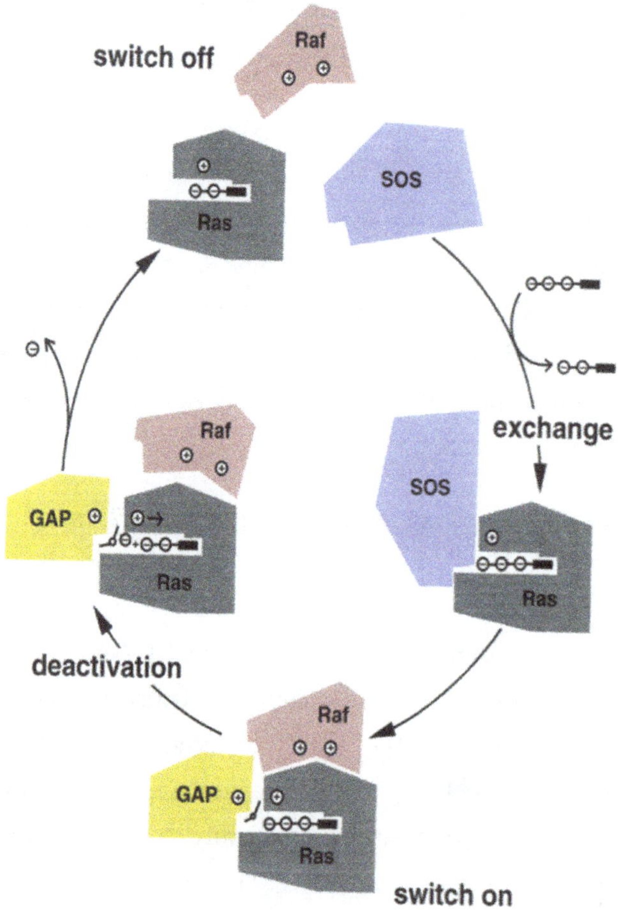

Fig. XIII.2 A more detailed description of Ras activation and deactivation. Reproduced with permission[303].

puzzling problem being the role of Gln61. Mutations of this residue in the RasGAP complex reduce the k_{cat} of the GTPase reaction by factors of 10^{-3}–10^{-7}, depending on the specific mutation. This large effect seems to be ubiquitous to different GAPs, including the Ran-RanGAP system, which does not operate through an arginine finger, but still contains a catalytically crucial glutamine (427). A related glutamine is also found in other G proteins (e.g., Q204 and Q227 in $G_{i\alpha 1}$ and $G_{s\alpha 1}$ respectively[414]). It is therefore clear that the proper positioning of Gln61 is crucial for effective catalysis. However, the nature of this positioning and its role in stabilizing the transition state are entirely unclear. This means that one of the most crucial links between structure and catalysis in Ras (and other signal transduction proteins) is of great interest.

In principle, there are several direct ways in which Gln61 could be contributing to the catalytic effect. That is, it could act "chemically" as a general base, it could provide direct electrostatic stabilization to the transition state, and it could also perhaps act by applying entropic or steric effects (see below). However, it is also equally possible that this residue is acting in an *indirect* way, the precise nature of which is not clear. Structural studies, mutational experiments and the use of transition state analogs have proven unable to elucidate the origin of the catalytic role of Gln61. In fact, all early works (such as, e.g., Refs. 452, 453) concluded that Gln must be crucial as a general base that extracts

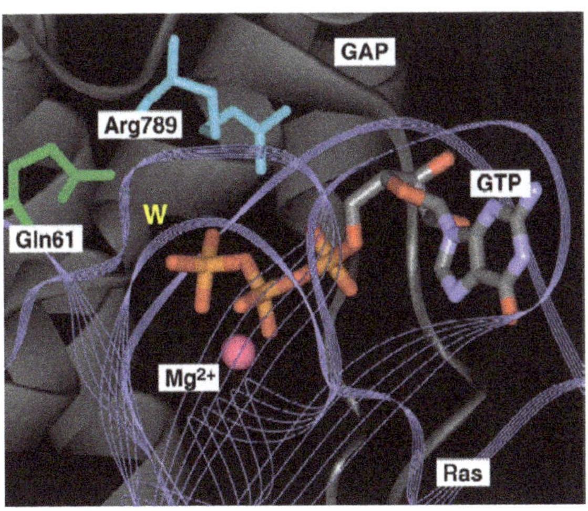

Fig. XIII.3 The GAP/Ras active site.

a proton from the nucleophilic water molecule. This view was drastically changed by our study[454], in which we demonstrated that Gln61 cannot be the general base in Ras (since the energy of the proposed proton transfer step in Ras is higher than the actual observed barrier for the GTPase reaction in solution). This (and subsequent) studies led to the conclusion that the GTP itself must be the elusive base, and therefore to the proposal of the GTP-as-base (or "phosphate-as-base") mechanism[455,456].

Despite some problematic arguments (see Ref. 457), it was our 1992 study[454] and the subsequent study of Schultz and coworkers[458] that established that Gln61 *cannot* be the general base. Next, we established the fact that the phosphate is the actual and eventual proton acceptor[456] and this finding is now widely accepted (see, e.g., Refs. 413, 447, 459, 460, amongst others). The basic question here is what the fate of the proton in the reaction of Ras is, and, since it is now agreed that the proton moves to the phosphate and not to Glu61, the question is reduced to the secondary question about the details of the exact pathway, and *not* to whether or not the phosphate accepts this proton. The mechanistic question could not be addressed by any of the available experiments and thus, following our initial study[454], we explored this issue repeatedly, eventually supplementing our earlier studies and quantum mechanical studies of the solution reaction by a detailed QM/MM study of the reaction surface for Ras[461], obtaining a transition state that is highly associative in character (P-O distance of 2.0 and 2.1 Å to the nucleophile and leaving group respectively), with concomitant proton transfer to a non-bridging oxygen on the phosphate.

It was found that the actual catalysis comes from the enzyme preorganization, including the electrostatic effect of the Mg^{2+} ion, and *not* from the specific mechanism used (the same mechanism used by the enzyme is much slower than in water[196]). Furthermore, the above assertion also overlooks the fact that the catalytic effect of OH^- cannot be determined experimentally (since we cannot have a protonated phosphate at the pH needed to study the OH^- effect; see discussion in Refs. 416, 462).

The realization that Gln cannot be the base led us to examine alternative mechanistic options, where the GTP accepts the proton in either a stepwise or concerted fashion (relative to the actual phosphoryl transfer), and is then hydrolyzed in either an associative or dissociative pathway (see, e.g., Ref. 222). The relevant mechanistic considerations ranged from an examination of linear free energy relationships[463] to considerations of the effect of the arginine finger on the associative and dissociative mechanisms[200]. In fact, the realization that the central mechanistic questions can only be resolved by theoretical studies (which must of course reproduce the relevant experimental observables) formed the basis for most of our mechanistic studies. The surface for a single water attack mechanism of the GTPase reaction in the RasGAP system was evaluated by *ab initio* QM/MM[461]. It was found that the reaction path preferentially proceeds through a concerted associative pathway, and that it is hard to experimentally establish whether this path is associative or dissociative as both transition states give rise to similar experimental observables. Interestingly, the new surfaces for the reaction in solution and in Ras involved an associative transition state, where the proton transfer from the attacking water is rather late[461]. The formation of an early, ktrt bond with the attacking water molecule was first observed by Refs. 464, 465, which considered it, however, to be a dissociative mechanism, and described the resulting

$H_2OPO_3^-$ (or $H_2PO_4^-$) fragment incorrectly as a metaphosphate, overlooking the fact that the nucleophilic attack occurs before the cleavage of the P-O bond.

The conclusion about the associative nature of the surface continued to be supported by more recent studies that pointed out, however, to an additional complication, namely the availability of a mechanism that involves a 2W molecule. That is, our most recent and most reliable study[466] focused on *ab initio* QM/MM examinations of the reference reaction in water. The reaction involves one nucleophilic water (1W) or a subsequent proton transfer through another water molecule (2W). This issue was investigated in several of our works[383,467,468] that also explored the validity of other works which did not examine the 1W path. Our studies, and in particular Refs. 466 and 468, examined by very careful 2D free energy surface for 1W and 2W. This study established that the 2W path has a lower barrier of at least 6 kcal/mol than that of the 1W path. For other studies see Ref. 457.

Using these findings, we calibrated empirical valence bond (EVB) surfaces on the refined *ab initio* solution surfaces and moved back to the studies of RasGAP and EF-Tu. In the RasGAP system discussed above, the main question was what the role of the oncogenic Gln 61 is, examining whether this residue is involved in chemical catalysis. Our study established that although RasGAP uses the 2W mechanism with either additional water or Gln 61, our early conclusion are still unchanged, since the arrival to the plateau, obtained during the nucleophilic attack of the water molecule, determines the rate (the barrier of the subsequent proton transfer is less than 1 kcal/mol higher than the plateau (see Ref. 466 and Fig. XIII.4)). The energy of the plateau appeared to be controlled by the electrostatic stabilization of the protein

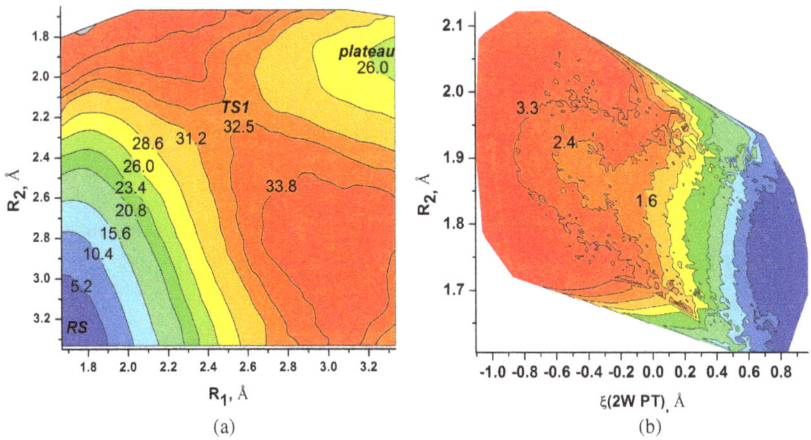

Fig. XIII.4 (a) The free energy surface in the R_1, R_2 space for the first step of the phosphate monoester hydrolysis (the cleavage of the P-O bond) in solution. (b) The surface for the 2W proton transfer step, showing that this step occurs spontaneously near transition state 1. Reproduced with permission[466].

and modulated by allosteric effects. We also established that the role of Gln 61 (in addition to its effect in electrostatic stabilization) may be in shuttling protons in the transition state, but this does not help the catalytic effect, since the proton transfer is not rate limiting (Fig. XIII.4).

As a crucial part of our analysis of the activation of the RasGAP system, we examined the effect of mutations of Arg789 (the so-called "arginine finger" depicted in Fig. XIII.5[75]) and Gln61 (the alleged general base). Since electrostatic effects are usually the leading contribution in catalysis[196], we first considered the simple approach of mutating the two residues to their nonpolar analogues, where all the sidechain atoms have zero residual charges. This type of analysis allows us to quantitatively determine what the electrostatic effect of the residues under consideration actually is. The results of the corresponding simulations are presented in Fig. XIII.5. It is seen from the figure that the observed effect of Arg789 is reproduced using the nonpolar analogue mutation

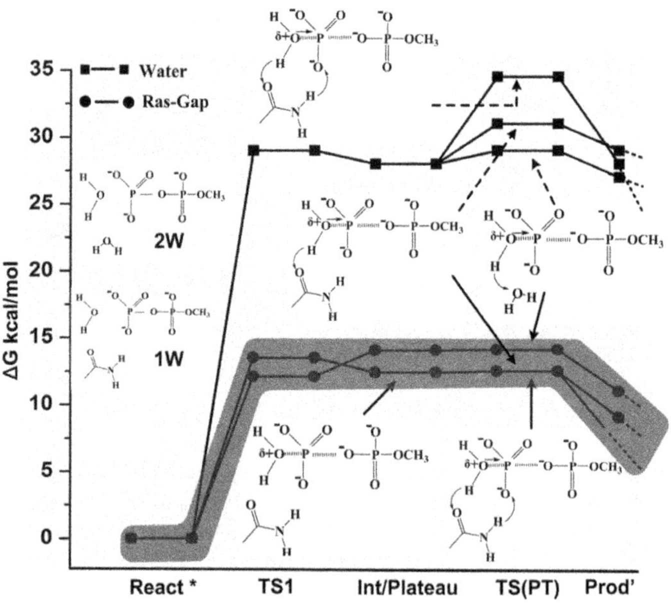

Fig. XIII.5 The EVB free energy profiles for the RasGAP reaction. The figure provides the profiles for different feasible paths as well as the barrier of the corresponding reference reactions in water. Reproduced with permission[466].

technique. This provides a clear demonstration of the fact that the main role of the arginine finger is to provide direct electrostatic stabilization of the transition state[224].

In examining the electrostatic effect in the activation process we considered the charge-charge in the reacting system during overall catalytic effect of GAP, focusing on its indirect effect through its interaction with Ras. A comparison of the X-ray structure of the isolated Ras to the structure of Ras in the RasGAP (henceforth referred to as Ras' for simplicity) reveals some structural change induced by the binding of GAP to Ras (such as the fact that Gln61 moves closer to Pγ[451]). However, although this comparison is quite instructive, it cannot tell us what the exact *functional* implications of these changes are. In particular, since the enzyme is flexible, changes in the positions of various residues do not necessarily reflect changes in the overall activation barrier. Furthermore, although transition state analogues can provide useful information, it is hard to deduce the exact relationship between transition state analogues and the corresponding transition state[469]. Therefore, we used linear response approximation[272] to determine the individual contributions of Ras in the RasGAP complex[224], and compared the catalytic effect of the isolated Ras to the contribution of Ras to the RasGAP complex. This was achieved by first evaluating the protein-substrate electrostatic interactions in the resting and transition states, and then subsequently decomposing them to the individual contributions from each residue. This study demonstrated that the structural reorganization between the free Ras and Ras' is indeed functionally important. Significant changes occur in the P-loop, Switch I residues (in particular Asp33 and Thr35), residues in the β_3 region (particularly Asp57), L4 in Switch II (particularly Gln62) and the Mg^{2+} ion. Therefore, identifying specific structural (rather than energetic) changes is not simple, since we obtain a similar trend with different starting configurations. Furthermore, our focus here is on average *energies* rather than average *structures*.

Apparently, as was found in our study[224], the interaction between GAP and Ras upon complexation moves Ras (which is part of the complex) from its stable resting state arrangement towards a configuration that stabilizes the transition state and the product. The corresponding catalytic effect is therefore indirect, and thus quite different from the direct effect of the arginine finger. It is important to take into account in this respect previous studies (e.g., Ref. 177), which proposed that the GAP-induced transition pushes the GTP-bound form of Ras towards the structure of the GDP-bound form of Ras.

Another source of support for our finding that Gln61 is acting in an indirect way comes from studies of the effect of mutations of $G_{s\alpha}$. The Q227L mutation (which is thought to have a similar role to that of the Q61L mutation in RasGAP) only leads to a reduction in k_{cat} of ~10^2 (see Refs. 470, 471) as compared to the 10^6 effect of the Q61L mutation. If the catalytic Gln were involved in a special direct interaction with the transition state, one would have expected similar contributions in the two related systems. Further support has been provided by a mutational experiment with an unnatural amino acid[147] substitution of Gln61 (Q61NGln) that also leads to a very small anti-catalytic effect in the RasGAP complex. This finding was also reproduced by our simulation studies.

In light of the above findings, we consider the indirect effect of GAP to be a major factor in the activation of the signal transduction switch. Apparently, relatively small structural rearrangements in Switch I, II, β_3 and the P-loop are sufficient to drastically change the transition state stabilization by the RasGAP complex. The system can be considered to be metastable with regard to its native sequence, where mutations of some residues change the energy landscape, leading to the destruction of the catalytic effect. Mutations of Gln61 are particularly effective at destroying the catalytic configuration (see Fig. XIII.6), leaving the system in its ON state, and thus causing cancer.

At this point it might be useful to comment about experimental analysis of the infrared absorption bands of the different phosphate groups (measured using Fourier transform infrared spectroscopy) in the isolated GTP and in Ras-bound GTP, pointing towards a large decrease in the P-O bond order, and the fact that the oxygen atoms of the β-phosphate interact more strongly with the protein environment

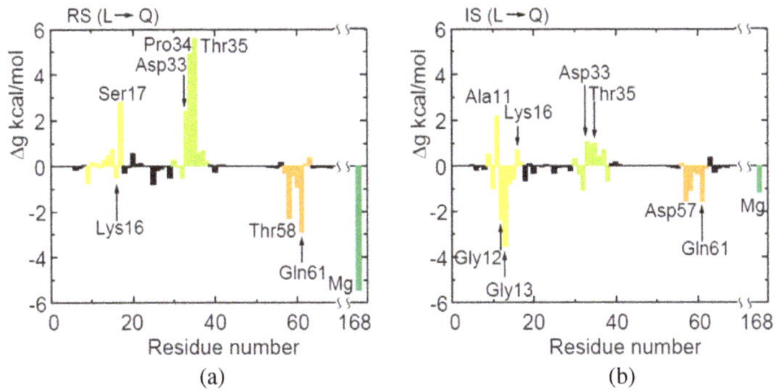

Fig. XIII.6 The main changes in the electrostatic contributions from the Ras residues due to the Q61L mutation. The changes in the resting state are compared to those in the transition state in (a) and (b), respectively. Negative and positive values correspond, respectively, to stabilization and destabilization due to the mutation. The changes in the key areas are highlighted: P-loop (yellow), Switch I (light green), parts of β_3 and L4 of the switch II (orange) and Mg (dark green). The calculated changes were obtained by the linear response approximation procedure and then scaled by a dielectric constant of 20 and 4 for the ionized and polar residues, respectively. Reproduced with permission[224].

than those of the γ-phosphate[472]. This effect becomes larger in the RasGAP complex[473]. Remarkably, this trend validates our early prediction[177] that there is a large shift of positive electrostatic potential towards the β-phosphate upon the binding of GAP[177,222] (Fig. XIII.7).

At any rate, the spectral shifts are indicative of a large change in the polarization of the substrate by its surrounding environment, which is the signature of the activation by Gap.

XIII.2 Elongation Factors and Ribosomal Activation

The activation of EF-Tu by the ribosome is closely related to the activity of Ras[469,474–477]. That is, the elongation cycle of protein synthesis uses the elongation factor (EF-Tu) complexes with GTP in order to deliver aminoacyl-tRNA to the A site of the ribosome. When the codon-anticodon

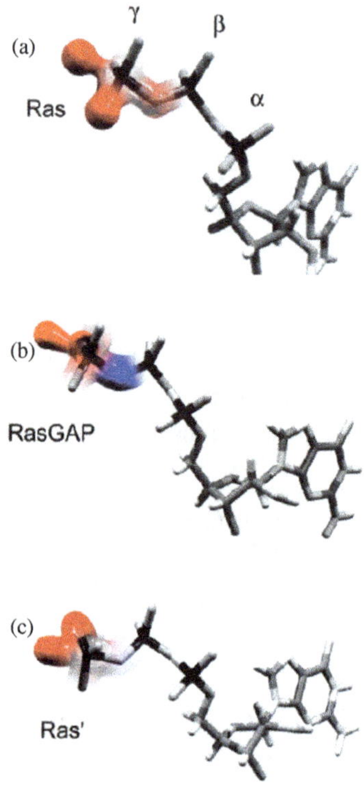

Fig. XIII.7 Change in electrostatic interactions between the protein residues and the γ- and β-phosphates of GTP, for the reaction catalyzed by (a) Ras, (b) RasGAP and (c) Ras'. Blue indicates a stabilizing shift and red indicates a destabilizing shift in the interactions, with the degree of stabilization or destabilization being proportional to the intensity of the color. Note that what is presented here is *not* the change in the electrostatic potential, but rather the actual change in electrostatic energy. Reproduced with permission[222].

interaction is cognate, the GTPase activity of EF-Tu is drastically increased. The conformational change following GTP hydrolysis that generates GDP and inorganic phosphate P_i leads to the dissociation of EF-Tu from the ribosome. The GTPase reaction in this system is stimulated by the ribosome but is slowed down by six orders of magnitude upon mutation of His 84 to Ala[469]. Exploiting the elucidation of the structure of the ribosome[477–488] and in particular the recent structural elucidation of the EF-Tu/ribosome complex[477–486,488], we explored the catalytic effect of the binding to the ribosome. First, we established that His 84 (that is pushed toward the GTP upon activation) is not a general base, since the proton transfer barrier was too high and the calculated pK_a was found to be above 10 (in agreement with the observed pH dependence[469]). Next we reproduced the catalytic effect with a 1W mechanism, showing that the activation involves major allosteric effects[489] in both the P loop and the His region.

After reevaluation of the reference solution reaction and establishment of the 2W mechanism, we revisited the EF-Tu system[466]. Our study established again that the His is not a general base and suggested that a 2W mechanism is more likely than our previous 1W mechanism. We also established again that we have a major allosteric effect. Interestingly, recent work[490,491] agreed with our conclusions on the RasGAP system but argued that EF-Tu has a special electrostatic stabilization that stabilizes the OH^- in the 1W mechanism and thus is not operating by the 2W mechanism. However, in our hands the 1W mechanism has too high energy.

Further progress in clarifying the catalytic effect of EF-Tu and EF-G was reported in Ref. 492. This work further established that the catalytic effect involves allosteric effects.

XIII.3 Related Issues and Systems

Other very important (and potentially instructive) benchmarks for the understanding of G-proteins are provided by transducin[493–495] and $G_{i\alpha 1}$ and $G_{s\alpha}$[414,496]. Here, it is important to understand what the main catalytic factors are, and how they are similar to or differ from those that are in operation in RasGAP. Both of these important signal transduction proteins have their catalytic glutamines in the same orientation and position as in the RasGAP complex, and both glutamines are thought to play a similar role as they do in RasGAP. Additionally, there exist other G-protein complexes such as the RapGAP[497] complex, which are capable of catalyzing GTP hydrolysis much like the RasGAP complex, yet both the catalytic Glu and Arg residues of the RasGAP complex are missing. This highlights the problem of oversimplifying the system to look at individual residues rather than the totality of the catalytic and allosteric effects.

XIII.4 Ras as an Example of the Role of Phosphorylation in Signaling

When the GTP substrate of Ras hydrolyzes to GDP, this results in a drop of the affinity of Ras towards its effector Raf which, in turn, leads to a dissociation of the active growth complex, thereby terminating the cell growth signal[498]. The nature of this control is discussed below.

Structural studies have provided crucial insight into the nature of these systems and the detailed protein/protein contacts that lead to signal transduction. For example, relevant developments in this area include the elucidation of the structure of RapA1-Raf[438], showing a complex of Raf with Rap which, for all practical considerations of the binding domain, is identical to Ras (see Fig. XIII.8). Furthermore, the structure of the Ras-binding domain of RalGEF, another effector of Ras, was solved by NMR spectroscopy[499], in addition to a variant of Ral-FEL called Rlf[499]. Recent years have seen a flurry of work in this area, so we would just like to mention a few more structural studies, including

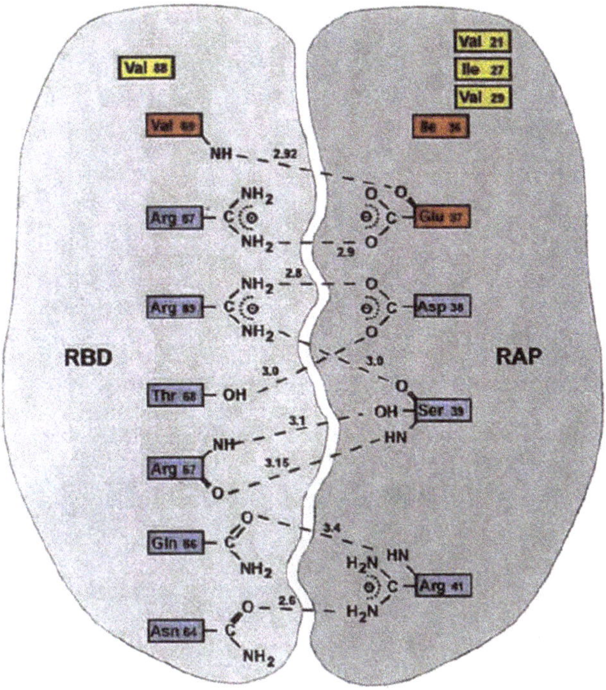

Fig. XIII.8 Ionized residues involved in the Ras/Rap interaction. Reproduced with permission[438].

combined experimental/computational attempts to understand the Ras-GDP interactions with mutants of Raf[500], as well as structural studies of the Rap-RapGAP complex[497], which is interesting due to its ability to perform GTP hydrolysis without the catalytic glutamine and arginine residues of RasGAP.

In any case, the availability of structural information for the RapA1/Raf complex allowed us to explore the key interactions between the two proteins[501]. Our careful study, which involved different levels of approximations, indicated that electrostatic interactions play a key role in protein/protein interactions. In particular, the calculations reproduced the observed effect of key mutations using a simple electrostatic model with a large effective dielectric constant.

Our ability to explore this system offers the option to develop drugs capable of mimicking one of the two partners, and thus breaking the signal transduction in monogenic Ras/Raf pairs. This idea is illustrated in Ref. 438.

The overall mechanism by which the interaction of Ras and Raf is coupled to the conversion of GTP to GTP can be explained by the simple electrostatic consideration described in Fig. XIII.9.

Another interesting and relevant example is provided by the GEFs such as CDC25[502] yeast or SOS in mammalian cells[503,504]. Exchange factors convert p21-GDP to p21-GTP by drastically reducing the affinity of both GDP and GTP in the ternary GEF-p21ras-GXP complex. Thus, a nucleotide free binary GEP-p21ras complex is formed that can bind GTP (the concentration of which in the cell is larger than that of GDP). This process 'activates' p21ras by generating the p21-GTP complex.

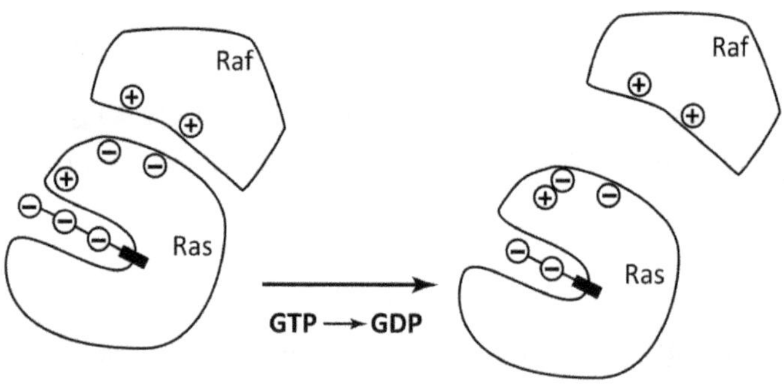

Fig. XIII.9 Nature of the transfer of the signal from GTP to GDP change to the Ras/Raf interaction. The change from GTP to GDP involves a loss of a negative charge. This reduction of negative charge is transmitted to the Ras surface as a reduction of the Ras/Raf electrostatic interaction and thus to dissociation of Rap. Reproduced with permission[303].

Our study of the mode of GEF activation[177] was performed before the availability of structural information on the complex (which became subsequently available; for a review see, e.g., Ref. 100). In this work[177], we evaluated the contribution of different residues of Ras to the binding of GTP and GDP. We then established that a significant number of the residues with large electrostatic group contributions are those whose mutations interfere with the action of GEFs. Interestingly, in considering the effect of GEF we had to account for the observation that the GEF reduces the binding affinity of GDP and GTP to p21ras by a similar amount. In doing so we concluded that the action of GEF might involve an interaction with Ras that pushes Gly60 towards Gly12 and that the resulting shift of Gly12 and its neighboring residues leads to a reduction in the interaction between the P-loop and the nucleotide. More extensive studies with the new structural information would be instructive, and potentially useful for drug design (see, e.g., Ref. 505).

At this point it is useful to provide some personal background. In 1991 I learned about a major advance regarding the protein p21ras, which is tightly involved in signal transduction (see above). I heard that a group at the Max Planck Institute in Heidelberg made a major progress in solving the structure of this protein. I therefore travelled from a meeting in Europe for a short visit at this great research center, hoping to get the X-ray structure of p21ras. I was also aware of a fierce fight between the German group (Withinghofer, Pai and Goody) and Sung-Hou Kim from Berkeley about the priority in the structural determination. When I arrived I got a very nice presentation of the structure, but was told that they could not give me the coordinates. Thus, I contacted Kim who was happy to provide me with his coordinates. With Ralf Lange and Thomas Schweins, who came from Hanover in

Germany as MS students, we explored the nature of the catalytic mechanism of p21ras. At first we focused on the structural paper published by the German group, in which the proposed mechanism implied that an oncogenic residue, Gln 61, accepts a proton from a water molecule, in what can be called "Gln as a base mechanism". This proposal seemed to be problematic, since Gln is not such a good base. However, to show that this is the wrong mechanism in the protein, we had to consider the basicity inside the protein. Fortunately, our EVB method provided a very powerful way to explore this issue with rather straightforward, yet reliable, calculations. That is, we used the EVB to show that in the enzyme the barrier for the proton transfer from the catalytic water molecule to Gln 61 (to form the nucleophilic OH^-) is significantly higher than the observed barrier for the actual enzymatic reaction. This meant that the reaction cannot involve a proton transfer to Gln. Our work was published in 1992 in an accelerated publication in *Biochemistry*. At any rate, our mechanism was supported by experiment with an unnatural amino acid replacing Gln 61 (see above).

My start with p21ras evolved to a major effort in understanding the energetics of phosphate hydrolysis in solution and proteins (see above) and the elucidation of the catalytic effect in GTPases. These studies include a major effort in understanding the activation of key GTPases such as the RasGAP system and the EF-Tu/ribosome system (see above). In all of these cases we identified electrostatic effects as the major factor in the activation process.

Chapter XIV

Electrostatic Basis for the Action of Molecular Machines

XIV.1 Molecular Motors

Conversion of chemical energy to work and generation of directional motion by biomolecular motors are paramount among the basic characteristics of living systems. Such energy transduction and motility processes ensure cellular viability and intracellular transport in regular and neuronal cells. Our initial studies of such systems started with the analysis of the energetics of F_1 ATPase[223] (see, e.g., Refs. 506,507 for a related study). We also note that our QM/MM study of phosphate hydrolysis in solution indicates that the ATP hydrolysis also involves the 2W path, but we also believe that this is not the rate-determining step in the ATPase reaction (as in the case in GTPases). In this respect, it may be useful to mention here other QM/MM studies of ATP hydrolysis in proteins and in solution[508–512], some of which have not attempted to obtain quantitative results (e.g., Ref. 508), and others which have not included a sufficiently large part of the environment or involved inadequate sampling[303]. Obviously, obtaining quantitative results is very challenging and our calibrated empirical valence bond method seems to do a good job on this front (see below).

While we continue to keep in mind the need for converging QM/MM surfaces we turn our attention to the overall conversion of chemical energy to work using our coarse-grained model in increasingly sophisticated studies of the conversion of chemical energy to work by F_1-ATPase. We started by evaluating the electrostatic free energy landscape[508], finding that the system undergoes an approximately 80° γ rotation from the starting structure followed by a concerted α/β dimer conformational changes along with γ rotation in the range of 40°. Such an intermediate conformational barrier has not been reproduced by other studies based on the 3D structures in an unbiased way. Adding the energetics of the ligand binding and hydrolysis allowed us to obtain the overall conformational/chemical landscape, and we also elucidated the basis for the experimentally observed catalytic dwell at 80° for the F_1 system. In a subsequent study[513,514] we started to explore the origin of the torque produced by the γ subunit rotation and reproduced both the observed torque and identified the coupling needed to generate such a torque (see Fig. XIV.1).

Studies of the above coupling in F_1-ATPase have been reported by other workers using coarse-grained models[515,516], microscopic simulations[517,518] and phenomenological approaches[519]. Interesting coarse-grained studies of related systems were also reported[520–523]. While the studies of F_1-ATPase were insightful, they have not produced a quantitative landscape, which could reproduce the observed

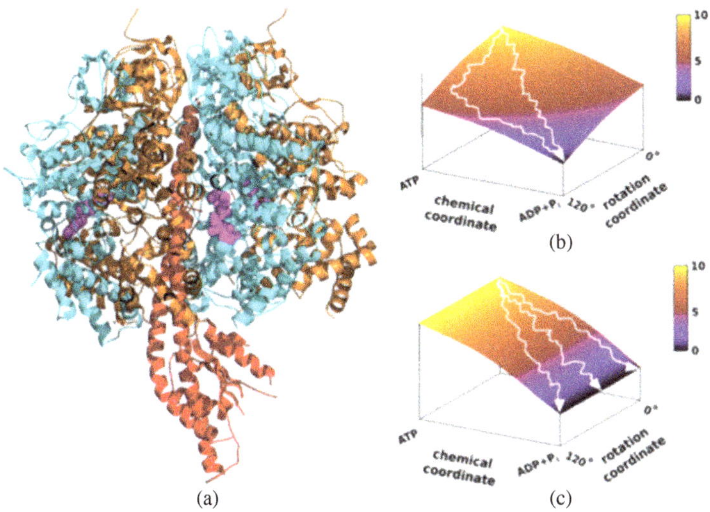

Fig. XIV.1 (a) The F_1-ATPase system, where the central stalk γ and subunits α and β are shown in red, orange and cyan, respectively. The relationship between the shape of the landscape and the generated torque is illustrated in (b) and (c), where the schematic landscapes are shown in terms of the chemical and rotational coordinate (omitting the barrier) to examine two extreme situations: the chemical free energy is (b) tightly coupled and (c) loosely coupled to the rotational coordinate. In (b) the chemical free energy decreases more when the rotational coordinate ϕ approaches 120° than when it approaches 0°. Hence the average trajectory ultimately moves from $\phi = 0°$ to $\phi = 120°$ and the system generates torque. In (c) we have the same reduction in the chemical free energy but the coupling to the rotational coordinate is not tight. So we obtain zero torque in the ϕ direction. Reproduced with permission[514].

directional rotation and the position of the catalytic dwell. Furthermore, in most cases it was proposed that the coupling reflects the steric effect between the stator and the rotor. One of the strongest arguments against the steric model is provided by interesting works[524,525], where it was established that the rotational motion occurs in the right direction even without most of the rotor subunit embedded inside the stator cavity. Significantly, this effect is reproduced by our electrostatic model (see Fig. XIV.2, and the supplementary information of Ref. 526 and Refs. 513, 514). It is unlikely to be reproduced by a consistent steric model, as most of the γ-subunits are now absent.

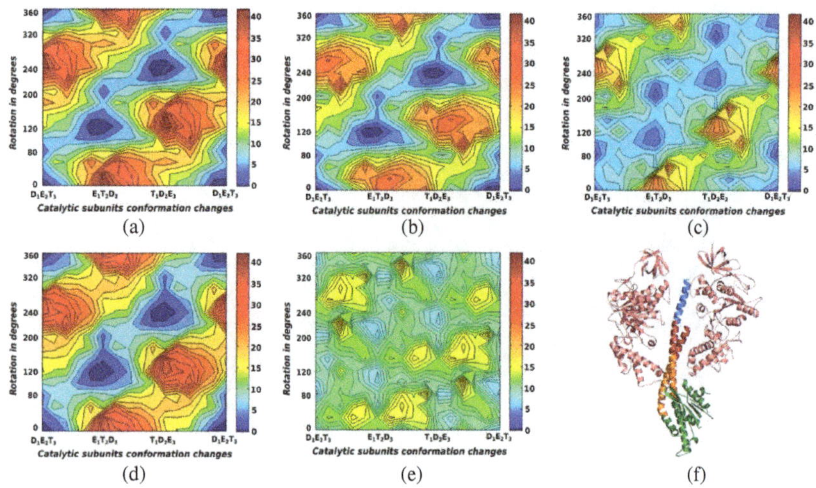

Fig. XIV.2 The coarse-grained electrostatic free energy surfaces for the γ-truncation systems without the addition of chemical free energy are shown for (a) $\gamma 1$, (b) $\gamma 2$, (c) $\gamma 3$, (d) $\gamma 4$, and (e) combined $\gamma 2$-$\gamma 3$. (f) The γ-subunit with a pair of opposing β-subunits are shown [Protein Data Bank ID: 1H8E], where $\gamma 1$, $\gamma 2$, $\gamma 3$, and $\gamma 4$ are colored in blue, red, orange, and green, respectively. Reproduced with permission[514].

Another aspect of the action of ATPase which is clearly electrostatic is the action of the F0-ATPase part of the ATPase where an Asp^- in a ring of Asp residues on the c-ring interacts with an Arg^+ residue on subunit-a in a way that is coupled to a proton transport. This coupling allows the pH gradient to rotate the F1 stalk and leads to ATP synthesis. This process has been simulated in our studies[527,528].

The study of the energy conversion in F_1-ATP synthases has been a part of the general attempt to understand the thermodynamic origin of directional motion in molecular machines. Our advance in this direction involved writing a general review that related our free energy landscape to the diagrams used to demonstrate the action of a Brownian ratchet[529]. Perhaps one of the most important findings was the possibility to project our landscape on the Brownian ratchet type diagram (see Fig. XIV.3) and hence identify the ratchet mechanism operating in the biological motors. This study emphasizes the importance of imposing microscopic reversibility on the forward and reverse processes that leads to a clear thermodynamic basis for the unidirectionality observed in biological machines.

At this point it may be useful to mention the beginning of my connection to the ATPase field. I met Peter Mitchell at a conference in 1978 and told him over breakfast about my electrostatic concepts. During this opportunity (which was just before he received the Nobel Prize)

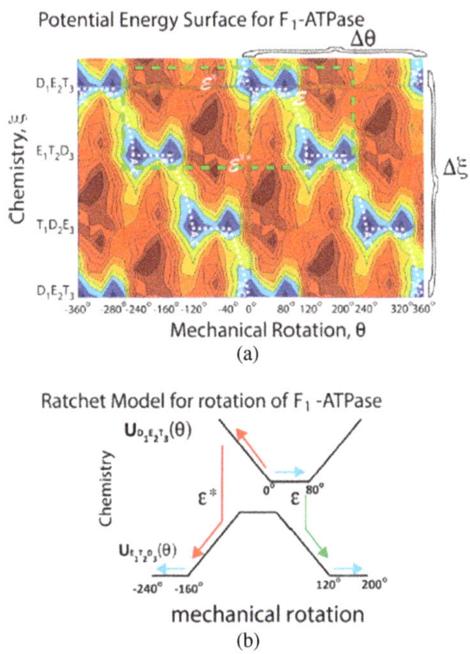

Fig. XIV.3 (a) Energy landscape for the F_1-ATPase from the work of Warshel and Mukherjee[530]. The fundamental periods $\Delta\xi$ and $\Delta\theta$ for which $U(\xi,\theta) = U(\xi \pm i\Delta\xi, \theta \pm j\Delta\theta)$ with $i, j \in \mathbb{Z}$ are shown. (b) A "ratchet" representation in terms of two 1D energy profiles with transitions between them for the two chemical states $D_1E_2T_3$ and $E_1T_2D_3$ from −240° to +200°. The remarkable and salient point is that the mechanism shown by the green arrows (clockwise rotation of 120°) seems by common sense to be far more likely than that shown by the dashed red arrows (counterclockwise rotation by 160°), but in fact, counter rotation is more likely than clockwise rotation. Reproduced with permission[529].

I learnt about his ideas on the key role of the electromotive force of the pH gradient. Subsequently, I tried to obtain more solid information about the energetics involved and the corresponding chemical barriers. Unfortunately, I found out that this information is largely missing from the available discussion and analysis. Although I generated a dedicated file for this project, I did not progress significantly until 2002, when I started to be involved in analyzing the energetics of ATPase, being armed with the information I gained from studies of phosphate hydrolysis in solutions and GTPases (see Chapter XIII). At this point I started to compile a table on the energetics of the different steps in the F1 motor and subsequently we started to explore the overall energy landscape using our coarse-grained model. Having the coarse-grained tool and the insight from careful compilation of the energetics of each step gave us a major advantage in the ability to gain molecular insight on the energy conversion process (see above). Similarly, our coarse-grained model allowed us to provide the first consistent structure-based molecular model for the conversion of pH gradient to activation of F_0-ATPase. This provided some connection to my initial discussion with Peter Mitchell.

Finally, we could ask the question of where the mechanochemical chemistry is actually stored. This cardinal question is clearly related to the electrostatic repulsion between the negative phosphate charges. However, the interesting point, that was identified in our work[531], is that a major part of the repulsion energy is released in water upon the separation of the already dissociated ADP-Pi complex rather than in the chemical step.

XIV.2 Myosin and Directionality

The action of myosin motors has major importance in biological systems and an understanding of the molecular basis of their function bears critical consequences in the field of medicine. Our initial work[532] and our subsequent studies (e.g., Ref. 533) on the myosin V molecular motor have explored the underlying link between the mechanochemical cycle that gives rise to its directional walking motion on the actin filament.

We found that the incorporation of the conformational free energy changes of the myosin lever angle leads to asymmetry that can favor the forward motion (however, the myosin motion in the backward directions was assumed to involve stretched arms rather than more complex motions). A subsequent study[533] that includes more degrees of freedom found that the backward motion can find a complex way to move with a similar barrier to that of the forward motion. The difference was found[534] to be in the ADP release step (see Fig. XIV.4). The fact that the directionality is determined by the activation barrier of the ADP release, and not by any

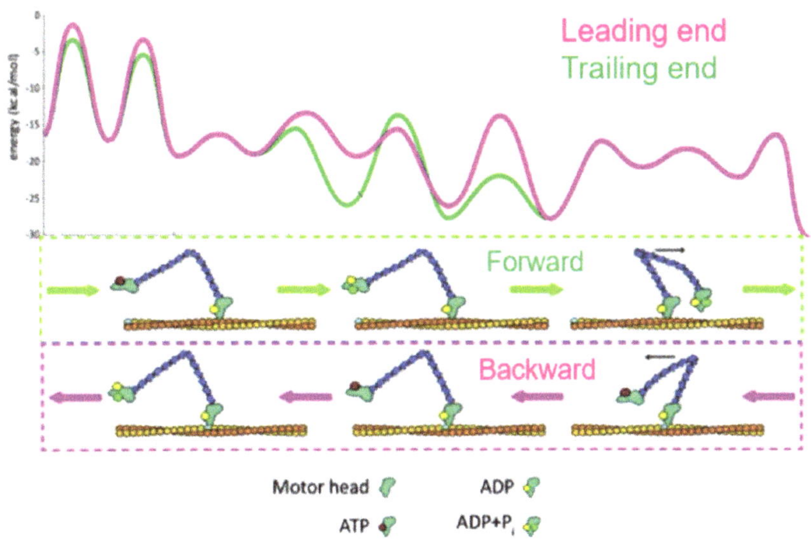

Fig. XIV.4 Free energy landscape for myosin stepping. The landscape for a forward step is shown in green, with the corresponding steps in the illustration at the bottom. The rate-determining step is the ADP release step, which is higher for the backward motion.

dynamical effect, has been recently confirmed[534] by careful evaluations of the activation barrier of the ADP release in myosin V and myosin VI.

XIV.3 The Ribosome and the Translocon

Our investigation of the ribosome and associated systems involved studies of protein insertion through the translocon[535,536], which was directly connected to the action of the ribosome in a study of the coupling between the ribosome and the translocon[535]. This study explored (by combining the empirical valence bond and coarse-grained models) the exciting finding[537] that a "stall" of the peptide synthesis process is released in special sequences and specific lengths by the translation through the translocon (this process is described in Fig. XIV.5). Remarkably we reproduced the trend in the experimental study by our modeling approach.

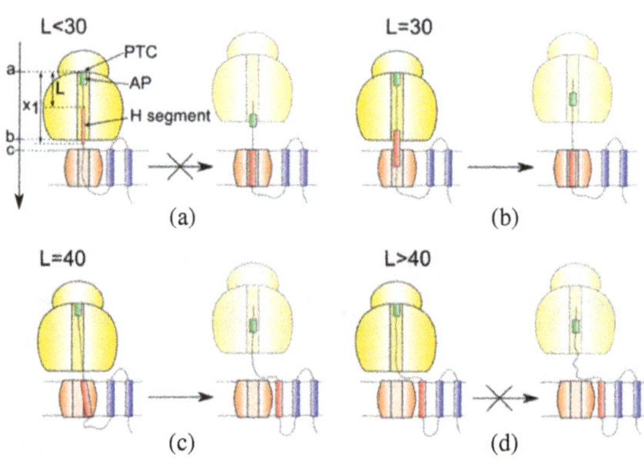

Fig. XIV.5 A schematic description of the effect of having different linker lengths (L) in the ribosome/translocon system. The linker starts from the pre-termination complex (PTC) and includes the arrested peptide (AP). (a) At short L the H segment insertion is blocked and the elongation is stalled. (b, c) At L = 30 and L = 40, the H segment insertion helps to break the stall and the ribosome is detached from the translocon (probably after completing the synthesis of the rest of the peptide). (d) Finally for L > 40, the H segment is inserted into the membrane, and it is not able to break the stalling. The figure also defines key parameters of the system. The arrow in (a) designates the x coordinate. Reproduced with permission[535].

Chapter XV

Some Electrostatic Factors in Bioenergetics

XV.1 Modeling the Pumping of Cytochrome c Oxidase

Cytochrome c oxidase (CcO) catalyzes the reduction of dioxygen to water, which is coupled to proton pumping across the membrane and the establishment of an electrochemical gradient that drives ATP synthesis. The mechanism of this reaction and its role as a proton pump has long been a major puzzle in bioenergetics[538–553]. The elucidation of the structure of CcO[381,539,554] (Fig. XV.1) and related systems (e.g., Refs. 555, 556) have provided an opportunity to explore a fundamental proton-pumping process. Unfortunately, despite the emergence of reasonable pictures from some studies[548,557–560], there are major difficulties in distinguishing between alternative conduction paths[547,561–565] and elucidating the factors that control proton pumping and gating in CcO.

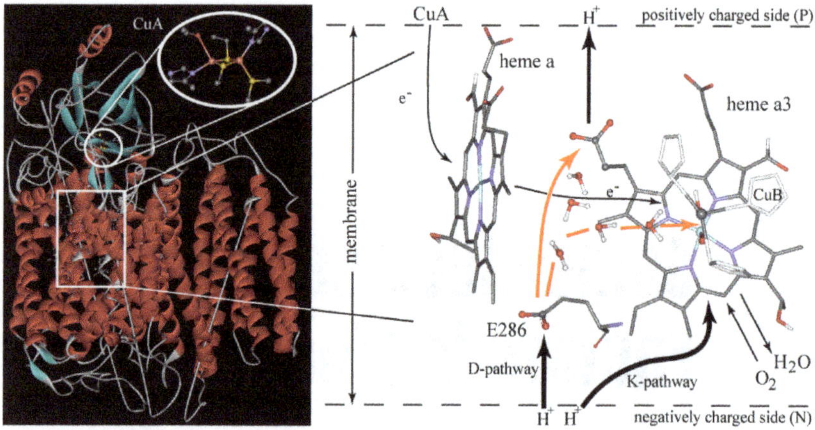

Fig. XV.1 A schematic description of the key elements in the COX systems. The figure considers heme a, heme a3, and two sides of the membrane, the entry and exit proton pathways. The pathways explored here are shown in orange on the right side of the diagram. Reproduced with permission[653].

This has remained one of the most challenging puzzles in bioenergetics, where it is essential to understand how the initial change in charge distribution pumps the protons against the pH gradient. The availability of structural, biochemical and mutational information gave hope for fast progress[538–550,557–560,566–580], but attempts to develop a detailed structure-based gating mechanism have led to disappointing results where almost always one ends up with a non-functional pump (e.g., Ref. 581). Many of the early works involved major problems in understanding the electrostatic control of proton transfer (e.g., Refs. 538,540) and many proposals involved Maxwell demons which contradicted macroscopic reversibility. Some workers in the field[538,540,564,565,582–588] have only considered the free energies of the different steps but have not yet evaluated the key activation barriers. Our works (e.g., Refs. 401, 589) led to very significant progress in exploring possible pumping cycles. However, the challenge of obtaining a pumping that is consistent with thermodynamic laws has not been overcome in a consistent way, but its solution must involve a proper electrostatic treatment.

XV.2 Transporters and Antiporters

Transport proteins play an important role in cell interactions and serve often as drug targets. As an example we consider our coarse-grained studies of such systems[590,591] that addressed the action of the bacterial Na^+/H^+ antiporter NhaA (whose action is described schematically in Fig. XV.2) and the MjNhaP1 transporter[590,591].

Our results (e.g., Fig. XV.3) should be applicable to the human homolog, involved in hypertension, kidney diseases[592], and cancer[593]. Another related problem appears in bacterial infections and cancer treatment, where multidrug resistance is becoming an alarming threat[594,595]. One component providing this resistance is multidrug efflux pumps. Thus, it should be interesting to explore the Multidrug and the Toxic compounds Extrusion (MATE) family in order to gain mechanistic understanding on how they function[596].

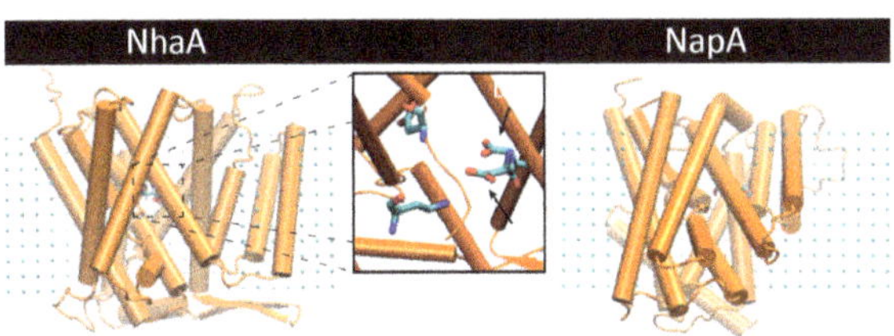

Fig. XV.2 Presentation of the Na^+/H^+ antiporter NhaA system used in our simulations. The protein appears in orange, and the membrane particle grid is shown as cyan spheres. The membrane has been truncated so it does not obscure the protein. Residues D133, D163, D164, and K300 are shown in a licorice presentation. Reproduced with permission[588].

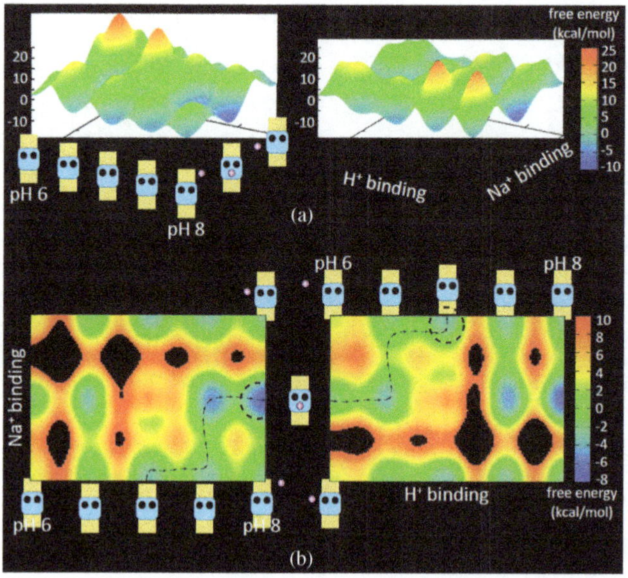

Fig. XV.3 The effective landscapes for the transporting function of NhaA. (a) A 3D surface of model 2. For simplicity the proton axis depicts two sequential proton bindings (see Results and Discussion, Reproducing and Understanding the Transport Function), as shown by the squares representing NhaA. The left and right surfaces represent the inward-facing and outward-facing states, respectively. (b) The same surfaces shown in 2D color. For visual clarity energies above 10 kcal/mol were truncated to black. Squares representing NhaA are shown to represent only some of the states. Dashed circles highlight the global minima. The arrows depict the path of a perfect transport cycle. Reproduced with permission[588].

Chapter XVI

G-Protein Coupled Receptors

The majority of cellular responses to hormones and neurotransmitters, as well as external agents like light and smell, are controlled by G-protein coupled receptors (GPCRs). The importance of this class of cellular receptors in understanding human diseases is emphasized through the fact that almost 50% of pharmaceutical drugs target GPCRs. The versatility of GPCR functionality lies in the fact that a single receptor molecule can be activated to elucidate different physiological responses mediated by different ligands or G-proteins. The extracellular domain of GPCRs can bind to agonists, antagonists or neutral agonists and translates this information to the cytoplasmic domain by activating, deactivating, or neutrally perturbing the action of heterotrimeric G-proteins, which leads to different downstream signaling events[597,598]. The spectacular advances in studies of the relevant complexes[599–603] give hope for advances in the detailed understanding of the activation process. Molecular dynamics simulation studies on GPCRs have provided interesting insights into the ligand binding and

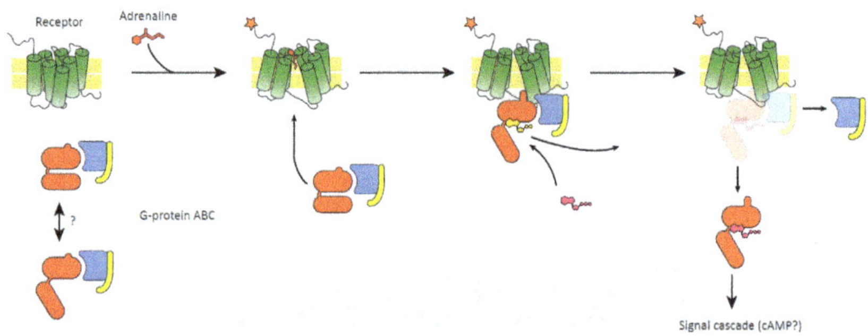

Fig. XVI.1 Schematic of the GPCR activation pathway. The receptor is activated upon binding of agonist (adrenaline in the case of the β2AR), while the G-protein is bound to GDP (indications for equilibrium between the conformations of free G-protein are mentioned in the main text). G-protein binds to the activated receptor, and undergoes the conformational changes required to facilitate exchange of GDP with GTP. G-protein bound to GTP then dissociates into the α-subunit and βγ-heterodimer, which elicit downstream signal transduction within the cell (e.g., increase or decrease of [cAMP]). Reproduced with permission[614].

allosteric activation[604–606]. An example of the activation cycle is given in Fig. XVI.1.

However, in spite of having excellent structural understanding of GPCR-mediated allosteric activation, we do not have reliable free energy landscapes for such processes. Here our coarse-grained model is poised to provide quantitative free energy surfaces that can lead to a thermodynamically robust understanding of the structure-function relationship (an example is provided in Fig. XVI.2). Our coarse-grained model has been used in exploring the landscape and functions of the β-2 adrenergic receptor[607], the GLP1R receptor[608] and the μ-opioid receptor[609] and other systems. Such studies can be refined by much more computer-intensive all-atom simulations (e.g., Ref. 610). It is obvious that further simulation studies will help in translating the emerging structural information to systematic information about the landscape of GPCRs and to further validate our findings and translate our insights to therapeutics (e.g., Ref. 611).

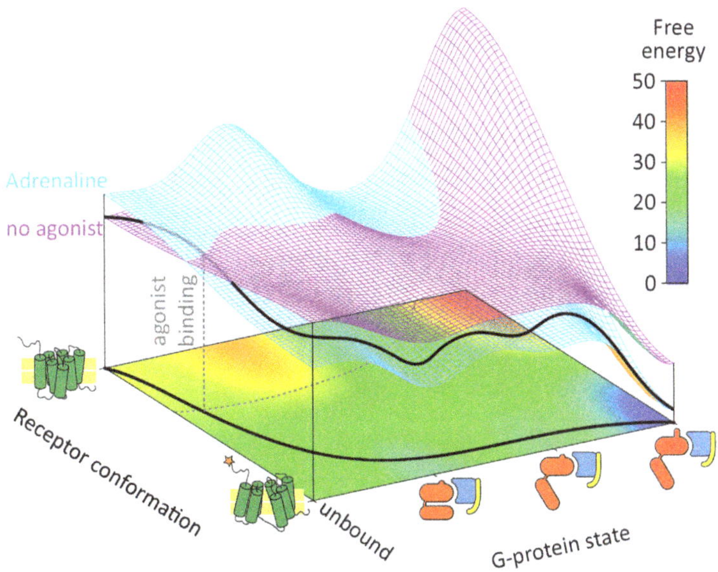

Fig. XVI.2 Three-dimensional semi-quantitative energy landscapes for the activation of the GPCR and G-protein system of Fig. XVI.1. The horizontal axes represent the conformation of the receptor and the state of the G protein, and the vertical z axis indicates the associated free energy. Two landscapes are presented, for the free energies with the agonist adrenaline and without it (in blue and purple, respectively). The G-protein state axis is constructed as a linear sequential coordinate to reduce the number of dimensions. Reproduced with permission[607].

With the emergence of more extensive computer power, it became possible to gain insights into the energetics of conformational changes in GPCRs. An instructive example is given by a recent study of the activation mechanism of S1PR1/3/5 by endogenous agonist S1P[610]. This study uses automated input-free path searching protocol and obtained reasonably reliable free energy landscape.

Interestingly, a major factor in the activation and control of GPCR is the change in the electrostatic contribution.

Chapter XVII

Other Systems

Many biological systems that were not discussed in previous chapters are controlled primarily by electrostatic effects. Below we will consider very briefly some examples.

XVII.1 Viruses

One of the most impressive examples of the role of electrostatic interactions is the assembly of the tobacco mosaic virus from its RNA and more than 2,000 protein subunits. One would have expected the helical protein coat to be assembled by sequential addition of the single subunits, but this is not what happens. They first aggregate into disks made up of two rings of 17 subunits each. On assembly into the virus, these rings undergo a transition from circular to helical symmetry with 16.34 subunits per turn. This transition happens on contact with the viral RNA but can also be mimicked by low pH. Structural analysis[111] has shown that the subunits are joined by two kinds of salt bridges: one bridge between neighboring subunits in the same ring

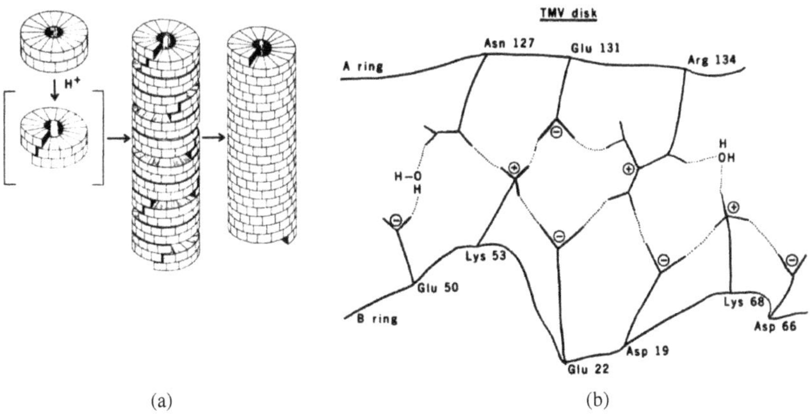

Fig. XVII.1 (a) Assembly of tobacco mosaic virus protein. Addition of H^+ to protein disks triggers their transition to "lock washers" and assembly into helical rods. (b) System of salt bridges between the two rings of the disk. Reproduced with permission[130].

that is preserved on transition from disk to virus — Arg 122 from the left radial helix of one subunit to Asp 88 of the right radial helix of its neighbors (Fig. XVII.1), and a network of salt bridges between the two rings of the disk that is lost in the course of that transition. The transition appears to be brought about by the deionization of two carboxyl groups and the formation of hydrogen bonds between them. Clearly the transition is controlled by electrostatic effects.

In terms of computational studies of viruses, the emphasis has been on the interaction of viruses with their targets in order to help in fighting viral diseases. An example is given by the current extensive study of the COVID-19 pandemic. These studies included our analysis of the binding interaction between the spike protein of the SARS-CoV-2 virus and the cell receptor ACE2[612]. This study tried to elucidate why SARS-CoV-2 binds more strongly to the ACE2 receptor than the original SARS-CoV virus. During this study we explored the effect of mutations on the binding process, by evaluating their electrostatic contribution. Our work that was finished already in May 2020 predicted the importance of the N501 mutation, which later was proved to be a key mutation in the newly spreading mutants. Subsequently[442], we explored the binding energy of the spike to the receptor and also the binding to the antibody that neutralizes the virus of some of the main emerging variants, and reproduced the observed trend. Subsequent studies include the effect of covalent inhibitors that block the action of the main protease of SARS-CoV-2 (e.g., Ref. 613).

XVII.2 Chromatin

Chromatin is a compact, folded higher-order structure that represses many DNA-dependent activities such as transcription, replication, and repair. The condensation of chromatin is therefore a prerequisite for normal cellular functions. At the first level of its organization, chromatin is highly uniform in most eukaryotes and can be considered as a linear array of nucleosomes that is formed by DNA and the histone octamer (HO). The HO comprises two copies each of the four core histone proteins: H2A, H2B, H3, and H4. The most regular part of the nucleosome, the NCP, includes 145–147 bp DNA wrapped as a left-handed 1.75 turn super helix around the HO (e.g. Ref. 614). The NCP is essentially a polyanion–polycation complex with a net charge of about −148e. Linker DNA having a variable length in the range 10–70 bp connects the NCPs to form an array of chromatin in the open, uncondensed "beads-on-a-string" state. Arrays of nucleosomes are folded, yielding a fiber of approximately 30 nm diameter (Fig. XVII.2).

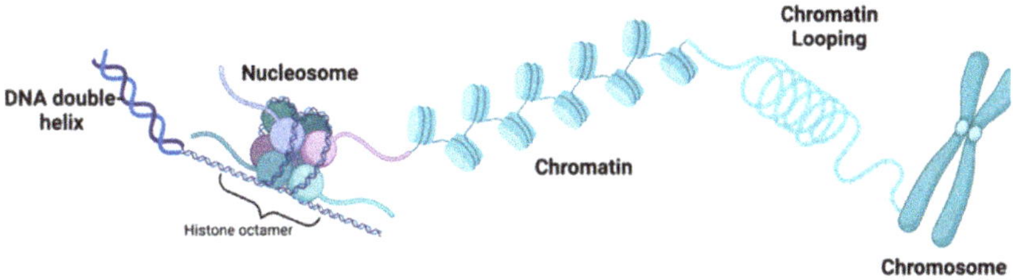

Fig. XVII.2 Nucleosome and chromatin structure. Chromatin is made up of nucleosomes, which consist of 147 bp of DNA wrapped around an octamer of histones. Each octamer consists of two molecules of each of the core histones, H2A, H2B, H3, and H4. H3 and H4 form a central tetramer $(H3\text{-}H4)_2$, which is flanked by H2A-H2B dimers. Nucleosomes are repeated structures that are compacted into chromatin. Increasingly condensed chromatin is organized into chromosomes. Reproduced with permission[654].

Considering that nucleosomes and chromatin are highly charged structures, it is not surprising that their assembly process is best described using electrostatically based coarse-grained modeling (e.g., Ref. 615).

XVII.3 Protein-Membrane Interactions

Membrane proteins play a major role in life processes[616] and thus there is great interest in realistic modeling of the functions of such systems. In some cases the functional properties can be simulated without taking the membrane into account, but in other cases it is crucial to consider the effect of the membrane in assessing the energetics of the action of membrane proteins (e.g., activation of ion channels[362,378,434,435], action of the translocon[535,620], F_0 ATPase[527]). Modeling the effect of the membrane can range from including a grid of polarizable dipoles to explicit all-atom simulations. However, moving to explicit all-atom membrane models is not always the most effective approach, in particular when one is interested in the landscape of processes that occur in long timescales and in large complex systems.

One of the most promising options is to use multiscale modeling and, in particular, coarse-grained (CG) models. Of course, CG models for proteins have been introduced long ago[41] and refined significantly in recent years[535,536,621–625]. Furthermore, CG models for membranes have been introduced and used extensively (e.g., Refs. 626–629) and there has been also some progress in CG models for protein insertion to membranes[630–633]. However, the focus on CG for the energetics of membrane proteins is more recent (e.g., Refs. 362, 434, 435, 514, 527, 530, 535, 536). The move in the direction of obtaining realistic CG models for membrane proteins has been slow in part because of the difficulties of obtaining unique points of calibration. That is, while in the case of CG for the energetics of proteins we have a relatively extensive benchmark of folding experiments[352], the situation with regard to the membrane part is more complex[634]. Here there are very few results that can be used with certainty in calibrating the energetics of peptide insertion (20-mer poly-Ala helix[635,636], 12-mer poly-Leu helix[637], the 23-amino acid M2δ segment of the nicotinic acetylcholine receptor[630,638]), and even the energy of charge and peptide insertion has been controversial[134,171,374,535,536,620,639–644]. Of course, one can benefit from the advances in microscopic simulations of penetration of charges as well as polar and nonpolar groups to membranes (e.g., Refs. 640,645–650).

Our effort in including membrane-protein interactions in our CG modeling has been based on a comprehensive fitting of calculated and

Table XVII.1 Absolute folding free energies, ΔG_{fold}, for several β-barrel integral membrane proteins in CG membranes calculated using refined CG models 1 and 2 compared to experimental values. Reproduced with permission[651].

			CG calculations		
				ΔG_{fold} (CG)	
Protein (PDB code)	Picture	ΔG_{fold} (exp.)	W_{mem} (Å)	Model 1	Model 2
OmpLA (1QD5)	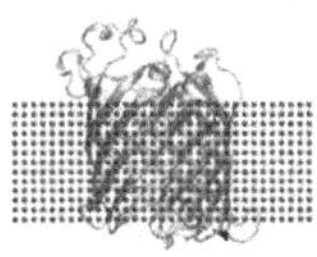	−32.45 (Ref. 63)	20	−34.94	−52.53
			24	−31.41	−33.01
OmpW (2F1V)	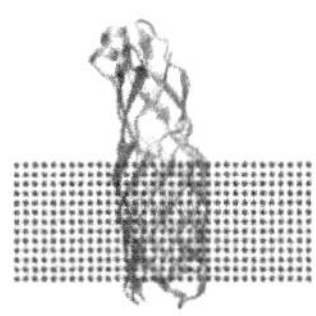	−18.60 (Ref. 64)	20	−4.66	−18.56
			24	−2.85	−14.29
PagP (1THQ)	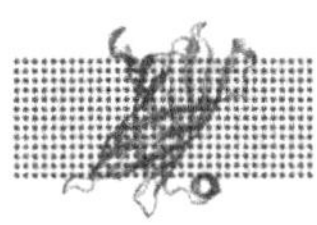	−24.40 (Ref. 64)	20	−51.49	−58.45
			24	−47.01	−37.62
OmpA (1QJP)		−3.40 (Ref. 71)	20	−26.17	−7.72
			24	−24.21	3.87
			28	−19.56	17.19

Note: All ΔG_{fold} are in kcal/mol and were computed using the formulation described in Ref. 651.

observed water/membrane partitioning energetics of amino acid side chains and model peptides. The resulting CG model[651] provides a reasonable agreement with experiment for absolute folding free energies of several β-barrel membrane proteins (Table XVII.1) as well as effects of point mutations on the relative stability for one of those proteins.

This CG model was used in several of our studies of the functions of membrane proteins, including F_0 ATPase, voltage-activated ion channels, and the translocon (see above). Again, as is pointed out in other parts of the book, the key factor is the balance between the energetics of the polar and nonpolar interactions. Here it is important to mention the effect of the membrane head groups, which should be considered in some cases (in most cases it can be done by using high effective dielectric).

Chapter XVIII

Concluding Remarks

The actions of living systems are controlled by cellular processes, and the understanding of such processes is the key for progress in medicine and related fields. With this issue in mind, it is crucial to ask what are the interactions that determine the nature of cellular processes and how to convert genetic and structural information into a functional picture.

This book considered the advancements in computer simulations, since the late 1960s, that led to the identification of electrostatic effects as the key to structure-function correlation in proteins. At this stage the field is moving to the next stage, where computers are 12 or more orders of magnitude faster than they were in the early seventies. Now, computers can run 1 microsecond simulation of 100,000 atoms in a day, whereas in the 1970s an IBM 360 could run 1 picosecond simulation of a very small protein in a few weeks. Hopefully, we are finally approaching the stage that millisecond simulations will become feasible and where we might start to obtain more converging results and probably accurate energies for some functional process. However, the studies conducted in the last 50 years yielded clear evidences that electrostatic energies provide the best correlation between structure and function and that many biological functions are controlled by the corresponding changes on electrostatic energies.

In the future, it is quite possible that artificial intelligence will provide powerful correlations between the sequence and function of biological systems. However, this may not provide a better understanding of the origin of the given correlation and such understanding is likely to be essential for rational advances. Here we believe that the focus on electrostatic factors will remain a key direction.

Finally, we may ask why nature relies on electrostatic energies more than van der Waals forces or hydrophobic interactions. The likely reason is that van der Waals forces decrease too rapidly with distance, making them ineffective for energy storage. Similarly, hydrophobic interactions do not respond adequately to changes in the charge distributions of reacting subsystems and are therefore not well-suited for storing significant energy. Overall, most biological energy transduction processes involve changes in charge, and the same holds true for biochemical transformations. This is why electrostatic interactions are so effective in controlling many biological processes.

References

1. Warshel, A.; Shalitin, Y. On the Interaction of Chymotrypsin with Ionized Substrate. *Isr. J. Chem.* **1966**, *4*, 85.
2. Lifson, S.; Warshel, A. Consistent Force Field for Calculations of Conformations, Vibrational Spectra, and Enthalpies of Cycloalkane and n-Alkane Molecules. *J. Chem. Phys.* **1968**, *49* (11), 5116–5129. https://doi.org/10.1063/1.1670007
3. Warshel, A. Calculations of Chemical Processes in Solutions. *J. Phys. Chem.* **1979**, *83* (12), 1640–1652. https://doi.org/10.1021/j100475a014
4. Sínanoĝlu, O.; Abdulnur, S. Hydrophobic Stacking of Bases and the Solvent Denaturation of DNA. *Photochem. Photobiol.* **1964**, *3* (4), 333–342. https://doi.org/10.1111/j.1751-1097.1964.tb08156.x
5. Claverie, P.; Pullman, B.; Caillet, J. Van Der Waals-London Interactions between Stacked Purines and Pyrimidines. *J. Theor. Biol.* **1966**, *12* (3), 419–434. https://doi.org/10.1016/0022-5193(66)90154-8
6. Huron, M. J.; Claverie, P. Calculation of the Interaction Energy of One Molecule with Its Whole Surrounding. I. Method and Application to Pure Nonpolar Compounds. *J. Phys. Chem.* **1972**, *76* (15), 2123–2133. https://doi.org/10.1021/j100659a011

7. Beveridge, D. L.; Kelly, M. M.; Radna, R. J. Theoretical Study of Solvent Effects on the Conformational Stability of Acetylcholine. *J. Am. Chem. Soc.* **1974**, *96* (12), 3769–3778. https://doi.org/10.1021/ja00819a010
8. Hylton, J.; Christoffersen, R. E.; Hall, G. G. A Model for the Ab Initio Calculation of Some Solvent Effects. *Chem. Phys. Lett.* **1974**, *26* (4), 501–504. https://doi.org/10.1016/0009-2614(74)80400-8
9. Pullman, A.; Pullman, B. New Paths in the Molecular Orbital Approach to Solvation of Biological Molecules. *Q. Rev. Biophys.* **1974**, *7* (4), 505–566. https://doi.org/10.1017/S0033583500001529
10. Warshel, A.; Levitt, M. Theoretical Studies of Enzymic Reactions: Dielectric, Electrostatic and Steric Stabilization of the Carbonium Ion in the Reaction of Lysozyme. *J. Mol. Biol.* **1976**, *103* (2), 227–249. https://doi.org/10.1016/0022-2836(76)90311-9
11. Schrödinger, E. Quantisierung Als Eigenwertproblem. *Ann. Phys.* **1926**, *384* (4), 361–376. https://doi.org/10.1002/andp.19263840404
12. Pople, J. A. Quantum Chemical Models (Nobel Lecture). *Angew. Chem. Int. Ed.* **1999**, *38* (13–14), 1894–1902. https://doi.org/10.1002/(SICI)1521-3773(19990712)38:13/14<1894::AID-ANIE1894>3.0.CO;2-H
13. Kohn, W. Nobel Lecture: Electronic Structure of Matter—Wave Functions and Density Functionals. *Rev. Mod. Phys.* **1999**, *71* (5), 1253–1266. https://doi.org/10.1103/RevModPhys.71.1253
14. Hendrickson, J. B. Molecular Geometry. V. Evaluation of Functions and Conformations of Medium Rings. *J. Am. Chem. Soc.* **1967**, *89* (26), 7036–7043. https://doi.org/10.1021/ja01002a036
15. Allinger, N. L.; Hirsch, J. A.; Miller, M. A.; Tyminski, I. J.; Van Catledge, F. A. Conformational Analysis. LX. Improved Calculations of the Structures and Energies of Hydrocarbons by the Westheimer Method. *J. Am. Chem. Soc.* **1968**, *90* (5), 1199–1210. https://doi.org/10.1021/ja01007a017

16. Levitt, M.; Lifson, S. Refinement of Protein Conformations Using a Macromolecular Energy Minimization Procedure. *J. Mol. Biol.* **1969**, *46* (2), 269–279. https://doi.org/10.1016/0022-2836(69)90421-5
17. Fletcher, R. *Practical Methods of Optimization*; John Wiley & Sons, 2000.
18. Warshel, A. *Computer Modeling of Chemical Reactions in Enzymes and Solutions*; John Wiley & Sons: New York, 1991.
19. Metropolis, N.; Rosenbluth, A. W.; Rosenbluth, M. N.; Teller, A. H.; Teller, E. Equation of State Calculations by Fast Computing Machines. *J. Chem. Phys.* **2004**, *21* (6), 1087–1092. https://doi.org/10.1063/1.1699114
20. Hwang, J. K.; King, G.; Creighton, S.; Warshel, A. Simulation of Free Energy Relationships and Dynamics of SN2 Reactions in Aqueous Solution. *J. Am. Chem. Soc.* **1988**, *110* (16), 5297–5311. https://doi.org/10.1021/ja00224a011
21. Schmitt, U. W.; Voth, G. A. Multistate Empirical Valence Bond Model for Proton Transport in Water. *J. Phys. Chem. B* **1998**, *102* (29), 5547–5551. https://doi.org/10.1021/jp9818131
22. Bentzien, J.; Muller, R. P.; Florián, J.; Warshel, A. Hybrid Ab Initio Quantum Mechanics/Molecular Mechanics Calculations of Free Energy Surfaces for Enzymatic Reactions: The Nucleophilic Attack in Subtilisin. *J. Phys. Chem. B* **1998**, *102* (12), 2293–2301. https://doi.org/10.1021/jp973480y
23. Villà, J.; Bentzien, J.; González-Lafont, À.; Lluch, J. M.; Bertran, J.; Warshel, A. Effective Way of Modeling Chemical Catalysis: Empirical Valence Bond Picture of Role of Solvent and Catalyst in Alkylation Reactions. *J. Comput. Chem.* **2000**, *21* (8), 607–625. https://doi.org/10.1002/(SICI)1096-987X(200006)21:8<607::AID-JCC3>3.0.CO;2-R
24. Muller, R. P.; Warshel, A. Ab Initio Calculations of Free Energy Barriers for Chemical Reactions in Solution. *J. Phys. Chem.* **1995**, *99* (49), 17516–17524. https://doi.org/10.1021/j100049a009

25. Åqvist, J.; Warshel, A. Simulation of Enzyme Reactions Using Valence Bond Force Fields and Other Hybrid Quantum/Classical Approaches. *Chem. Rev.* **1993**, *93* (7), 2523–2544. https://doi.org/10.1021/cr00023a010
26. Chang, Y. T.; Miller, W. H. An Empirical Valence Bond Model for Constructing Global Potential Energy Surfaces for Chemical Reactions of Polyatomic Molecular Systems. *J. Phys. Chem.* **1990**, *94* (15), 5884–5888. https://doi.org/10.1021/j100378a052
27. Vuilleumier, R.; Borgis, D. An Extended Empirical Valence Bond Model for Describing Proton Transfer in $H^+(H_2O)_n$ Clusters and Liquid Water. *Chem. Phys. Lett.* **1998**, *284* (1), 71–77. https://doi.org/10.1016/S0009-2614(97)01365-1
28. Shaik, S. S.; Schlegel, H. B.; Wolfe, S. *Theoretical Aspects of Physical Organic Chemistry: The SN2 Mechanism*; John Wiley & Sons: NY, 1992.
29. Wesolowski, T.; Muller, R. P.; Warshel, A. Ab Initio Frozen Density Functional Calculations of Proton Transfer Reactions in Solution. *J. Phys. Chem.* **1996**, *100* (38), 15444–15449. https://doi.org/10.1021/jp961068x
30. Hong, G.; Rosta, E.; Warshel, A. Using the Constrained DFT Approach in Generating Diabatic Surfaces and Off-Diagonal Empirical Valence Bond Terms for Modeling Reactions in Condensed Phases. *J. Phys. Chem. B* **2006**, *110* (39), 19570–19574. https://doi.org/10.1021/jp0625199
31. Hong, G.; Strajbl, M.; Wesolowski, T. A.; Warshel, A. Constraining the Electron Densities in DFT Method as an Effective Way for Ab Initio Studies of Metal-Catalyzed Reactions. *J. Comput. Chem.* **2000**, *21* (16), 1554–1561. https://doi.org/10.1002/1096-987X(200012)21:16<1554::AID-JCC12>3.0.CO;2-I
32. Olsson, M. H. M.; Hong, G.; Warshel, A. Frozen Density Functional Free Energy Simulations of Redox Proteins: Computational Studies of the Reduction Potential of Plastocyanin and Rusticyanin. *J. Am. Chem. Soc.* **2003**, *125* (17), 5025–5039. https://doi.org/10.1021/ja0212157

33. Štrajbl, M.; Hong, G.; Warshel, A. Ab Initio QM/MM Simulation with Proper Sampling: "First Principle" Calculations of the Free Energy of the Autodissociation of Water in Aqueous Solution. *J. Phys. Chem. B* **2002**, *106* (51), 13333–13343. https://doi.org/10.1021/jp021625h
34. Jacob, C. R.; Visscher, L. Calculation of Nuclear Magnetic Resonance Shieldings Using Frozen-Density Embedding. *J. Chem. Phys.* **2006**, *125* (19), 194104. https://doi.org/10.1063/1.2370947
35. Iannuzzi, M.; Kirchner, B.; Hutter, J. Density Functional Embedding for Molecular Systems. *Chem. Phys. Lett.* **2006**, *421* (1), 16–20. https://doi.org/10.1016/j.cplett.2005.08.155
36. Klüner, T.; Govind, N.; Wang, Y. A.; Carter, E. A. Prediction of Electronic Excited States of Adsorbates on Metal Surfaces from First Principles. *Phys. Rev. Lett.* **2001**, *86* (26), 5954–5954. https://doi.org/10.1103/PhysRevLett.86.5954
37. Huang, P.; Carter, E. A. Local Electronic Structure around a Single Kondo Impurity. *Nano Lett.* **2006**, *6* (6), 1146–1150. https://doi.org/10.1021/nl0602847
38. Neugebauer, J.; Baerends, E. J. Exploring the Ability of Frozen-Density Embedding to Model Induced Circular Dichroism. *J. Phys. Chem. A* **2006**, *110* (28), 8786–8796. https://doi.org/10.1021/jp0622280
39. Warshel, A. Computer Simulations of Enzymatic Reactions. *Curr. Opin. Struct. Biol.* **1992**, *2*, 230–236. https://doi.org/10.1016/0959-440X(92)90151-V
40. Warshel, A.; Parson, W. W. Dynamics of Biochemical and Biophysical Reactions: Insight from Computer Simulations. *Q. Rev. Biophys.* **2001**, *34* (4), 563–679. https://doi.org/10.1017/S0033583501003730
41. Levitt, M.; Warshel, A. Computer Simulation of Protein Folding. *Nature* **1975**, *253* (5494), 694–698. https://doi.org/10.1038/253694a0
42. Schlick, T. Modeling Superhelical DNA: Recent Analytical and Dynamic Approaches. *Curr. Opin. Struct. Biol.* **1995**, *5* (2), 245–262. https://doi.org/10.1016/0959-440X(95)80083-2

43. Papazyan, A.; Maroncelli, M. Rotational Dielectric Friction and Dipole Solvation: Tests of Theory Based on Simulations of Simple Model Solutions. *J. Chem. Phys.* **1995**, *102* (7), 2888–2919. https://doi.org/10.1063/1.468667
44. Dryga, A.; Warshel, A. Renormalizing SMD: The Renormalization Approach and Its Use in Long Time Simulations and Accelerated PMF Calculations of Macromolecules. *J. Phys. Chem. B* **2010**, *114* (39), 12720–12728. https://doi.org/10.1021/jp1056122
45. Olsson, M. H. M.; Warshel, A. Monte Carlo Simulations of Proton Pumps: On the Working Principles of the Biological Valve That Controls Proton Pumping in Cytochrome c Oxidase. *Proc. Natl. Acad. Sci. U.S.A.* **2006**, *103* (17), 6500–6505. https://doi.org/10.1073/pnas.0510860103
46. Jackson, J. D.; Fox, R. F. Classical Electrodynamics, 3rd Ed. *Am. J. Phys.* **1999**, *67* (9), 841–842. https://doi.org/10.1119/1.19136
47. Warshel, A.; Kato, M.; Pisliakov, A. V. Polarizable Force Fields: History, Test Cases, and Prospects. *J. Chem. Theory Comput.* **2007**, *3* (6), 2034–2045. https://doi.org/10.1021/ct700127w
48. Hopfinger, A. J. In *Peptides, Polypeptides and Proteins*; Blout, E. R., Bovey, F. A., Goodman, M., Lotan, N., Eds.; John Wiley & Sons: New York, 1974; pp 77–78.
49. Warshel, A. Calculations of Enzymatic Reactions: Calculations of pKa, Proton Transfer Reactions, and General Acid Catalysis Reactions in Enzymes. *Biochemistry* **1981**, *20* (11), 3167–3177. https://doi.org/10.1021/bi00514a028
50. Barnes, P.; Finney, J. L.; Nicholas, J. D.; Quinn, J. E. Cooperative Effects in Simulated Water. *Nature* **1979**, *282* (5738), 459–464. https://doi.org/10.1038/282459a0
51. Thole, B. T. Molecular Polarizabilities Calculated with a Modified Dipole Interaction. *Chem. Phys.* **1981**, *59* (3), 341–350. https://doi.org/10.1016/0301-0104(81)85176-2

52. Warshel, A.; Russell, S. T. Calculations of Electrostatic Interactions in Biological Systems and in Solutions. *Q. Rev. Biophys.* **1984**, *17* (3), 283–422. https://doi.org/10.1017/S0033583500005333
53. Onsager, L. Electric Moments of Molecules in Liquids. *J. Am. Chem. Soc.* **1936**, *58*, 1486–1493. https://doi.org/10.1021/ja01299a050
54. Kirkwood, J. G. The Dielectric Polarization of Polar Liquids. *J. Chem. Phys.* **2004**, 7 (10), 911–919. https://doi.org/10.1063/1.1750343
55. Jackson, J. D. *Classical Electrodynamics*, 3rd ed.; John Wiley & Sons: New York, 1999.
56. Böttcher, C. J. F. *Theory of Electric Polarization*, 2nd ed.; Elsevier: Amsterdam, 1973.
57. Dufner, H.; Kast, S. M.; Brickmann, J.; Schlenkrich, M. Ewald Summation versus Direct Summation of Shifted-Force Potentials for the Calculation of Electrostatic Interactions in Solids: A Quantitative Study. *J. Comput. Chem.* **1997**, *18* (5), 660–676. https://doi.org/10.1002/(SICI)1096-987X(19970415)18:5<660::AID-JCC7>3.0.CO;2-L
58. Figueirido, F.; Del Buono, G. S.; Levy, R. M. On Finite-Size Corrections to the Free Energy of Ionic Hydration. *J. Phys. Chem. B* **1997**, *101* (29), 5622–5623. https://doi.org/10.1021/jp970882x
59. Hummer, G.; Pratt, L. R.; García, A. E. Free Energy of Ionic Hydration. *J. Phys. Chem. B* **1996**, *100*, 1206–1215. https://doi.org/10.1021/jp951011v
60. Sham, Y. Y.; Warshel, A. The Surface Constraint All Atom Model Provides Size Independent Results in Calculations of Hydration Free Energies. *J. Chem. Phys.* **1998**, *109* (18), 7940–7944. https://doi.org/10.1063/1.477441
61. Hünenberger, P. H.; McCammon, J. A. Ewald Artifacts in Computer Simulations of Ionic Solvation and Ion–Ion Interaction: A Continuum Electrostatics Study. *J. Chem. Phys.* **1999**, *110* (4), 1856–1872. https://doi.org/10.1063/1.477873
62. Aaqvist, J. Comment on "Transferability of Ion Models." *J. Phys. Chem.* **1994**, *98*, 8253–8255. https://doi.org/10.1021/j100084a049

63. Åqvist, J.; Hansson, T. Analysis of Electrostatic Potential Truncation Schemes in Simulations of Polar Solvents. *J. Phys. Chem. B* **1998**, *102* (19), 3837–3840. https://doi.org/10.1021/jp973207w
64. Lee, F. S.; Warshel, A. A Local Reaction Field Method for Fast Evaluation of Long-range Electrostatic Interactions in Molecular Simulations. *J. Chem. Phys.* **1992**, *97* (5), 3100–3107. https://doi.org/10.1063/1.462997
65. King, G.; Warshel, A. A Surface Constrained All-atom Solvent Model for Effective Simulations of Polar Solutions. *J. Chem. Phys.* **1989**, *91* (6), 3647–3661. https://doi.org/10.1063/1.456845
66. Kirkwood, J. G. Theory of Solutions of Molecules Containing Widely Separated Charges with Special Application to Zwitterions. *J. Chem. Phys.* **2004**, *2* (7), 351–361. https://doi.org/10.1063/1.1749489
67. Lee, F. S.; Chu, Z. T.; Warshel, A. Microscopic and Semimicroscopic Calculations of Electrostatic Energies in Proteins by the POLARIS and ENZYMIX Programs. *J. Comput. Chem.* **1993**, *14* (2), 161–185. https://doi.org/10.1002/jcc.540140205
68. Alper, H.; Levy, R. M. Dielectric and Thermodynamic Response of a Generalized Reaction Field Model for Liquid State Simulations. *J. Chem. Phys.* **1993**, *99* (12), 9847–9852. https://doi.org/10.1063/1.465468
69. Warshel, A.; Levitt, M. Theoretical Studies of Enzymic Reactions:... *J. Mol. Biol.* **1976**, *103* (2), 227–249. https://doi.org/10.1016/0022-2836(76)90311-9
70. Stern, H. A.; Rittner, F.; Berne, B. J.; Friesner, R. A. Combined Fluctuating Charge and Polarizable Dipole Models: Application to a Five-Site Water Potential Function. *J. Chem. Phys.* **2001**, *115* (5), 2237–2251. https://doi.org/10.1063/1.1376165
71. Ren, P.; Ponder, J. W. Consistent Treatment of Inter- and Intramolecular Polarization in Molecular Mechanics Calculations. *J. Comput. Chem.* **2002**, *23* (16), 1497–1506. https://doi.org/10.1002/jcc.10127

72. Warshel, A.; Sussman, F.; King, G. Free energy of charges in solvated proteins: microscopic calculations using a reversible charging process. *(ACS Publ.)*. https://doi.org/10.1021/bi00374a006
73. Sham, Y. Y.; Chu, Z. T.; Warshel, A. Consistent Calculations of pKa's of Ionizable Residues in Proteins: Semi-Microscopic and Microscopic Approaches. *J. Phys. Chem. B* **1997**, *101* (22), 4458–4472. https://doi.org/10.1021/jp963412w
74. Del Buono, G. S.; Figueirido, F. E.; Levy, R. M. Intrinsic pKs of Ionizable Residues in Proteins: An Explicit Solvent Calculation for Lysozyme. *Proteins: Struct., Funct., Bioinf.* **1994**, *20* (1), 85–97. https://doi.org/10.1002/prot.340200109
75. Saito, M. Molecular dynamics/free energy study of a protein in solution with all degrees of freedom and long-range Coulomb interactions. *J. Phys. Chem.* **1995**, *99*, 17043–17048. https://doi.org/10.1021/j100046a033
76. Warshel, A.; Papazyan, A.; Muegge, I. Microscopic and Semimacroscopic Redox Calculations: What Can and Cannot Be Learned from Continuum Models. *JBIC* **1997**, 2 (1), 143–152. https://doi.org/10.1007/s007750050119
77. Warshel, A. Dynamics of reactions in polar solvents. Semiclassical trajectory studies of electron-transfer and proton-transfer reactions. *J. Phys. Chem.* **1982**, *86*, 2218–2224. https://doi.org/10.1021/j100209a016
78. Kato, M.; Warshel, A. Through the Channel and around the Channel: Validating and Comparing Microscopic Approaches for the Evaluation of Free Energy Profiles for Ion Penetration through Ion Channels. *J. Phys. Chem. B* **2005**, *109* (41), 19516–19522. https://doi.org/10.1021/jp053208l
79. Jarzynski, C. Nonequilibrium Equality for Free Energy Differences. *Phys. Rev. Lett.* **1997**, *78* (14), 2690–2693. https://doi.org/10.1103/PhysRevLett.78.2690

80. Jansen, A.; Bauer, P.; Howard, R.; Hess, B.; Lindahl, E. Constant-pH Molecular Dynamics Simulations of a Proton-Gated Ion Channel. *Eur. Biophys. J.* **2023**, *52* (Suppl 1), S86.
81. Oikonomou, E.; Gruber, T.; Chandra, A. R.; Höller, S.; Alzheimer, C.; Wellein, G.; Huth, T. 2D-Dwell-Time Analysis with Simulations of Ion-Channel Gating Using High-Performance Computing. *Biophys. J.* **2023**, *122* (7), 1287–1300. https://doi.org/10.1016/j.bpj.2023.02.023
82. Purcell, E. M. *Electricity and Magnetism*; McGraw-Hill: 1965.
83. Liu, Y.; Ichiye, T. The Static Dielectric Constant of the Soft Sticky Dipole Model of Liquid Water: Monte Carlo Simulation. *Chem. Phys. Lett.* **1996**, *256* (3), 334–340. https://doi.org/10.1016/0009-2614(96)00448-4
84. Coalson, R. D.; Duncan, A. Statistical Mechanics of a Multipolar Gas: A Lattice Field Theory Approach. *J. Phys. Chem.* **1996**, *100* (7), 2612–2620. https://doi.org/10.1021/jp952824m
85. Papazyan, A.; Warshel, A. Continuum and Dipole-Lattice Models of Solvation. *J. Phys. Chem. B* **1997**, *101* (51), 11254–11264. https://doi.org/10.1021/jp971632j
86. Papazyan, A.; Warshel, A. Effect of Solvent Discreteness on Solvation. *J. Phys. Chem. B* **1998**, *102* (27), 5348–5357. https://doi.org/10.1021/jp981725g
87. Russell, S. T.; Warshel, A. Calculations of Electrostatic Energies in Proteins: The Energetics of Ionized Groups in Bovine Pancreatic Trypsin Inhibitor. *J. Mol. Biol.* **1985**, *185* (2), 389–404. https://doi.org/10.1016/0022-2836(85)90411-5
88. HaDuong, T.; Phan, S.; Marchi, M.; Borgis, D. Electrostatics on Particles: Phenomenological and Orientational Density Functional Theory Approach. *J. Chem. Phys.* **2002**, *117* (2), 541–556. https://doi.org/10.1063/1.1481858
89. Schaefer, M.; Karplus, M. A Comprehensive Analytical Treatment of Continuum Electrostatics. *J. Phys. Chem.* **1996**, *100* (5), 1578–1599. https://doi.org/10.1021/jp9521621

90. Simonson, T. Electrostatics and Dynamics of Proteins. *Rep. Prog. Phys.* **2003**, *66* (5), 737. https://doi.org/10.1088/0034-4885/66/5/202
91. Warshel, A.; Naray-Szabo, G.; Sussman, F.; Hwang, J. K. How do serine proteases really work? *Biochemistry* **1989**, *28*, 3629–3637. https://doi.org/10.1021/bi00435a001
92. Sham, Y. Y.; Chu, Z. T.; Tao, H.; Warshel, A. Examining Methods for Calculations of Binding Free Energies: LRA, LIE, PDLD-LRA, and PDLD/S-LRA Calculations of Ligands Binding to an HIV Protease. *Proteins: Struct., Funct., Bioinf.* **2000**, *39* (4), 393–407. https://doi.org/10.1002/(SICI)1097-0134(20000601)39:4<393::AID-PROT120>3.0.CO;2-H
93. Schutz, C. N.; Warshel, A. What Are the Dielectric "Constants" of Proteins and How to Validate Electrostatic Models? *Proteins* **2001**, *44* (4), 400–417. https://doi.org/10.1002/prot.1106
94. Kollman, P. A.; Massova, I.; Reyes, C.; Kuhn, B.; Huo, S.; Chong, L.; Lee, M.; Lee, T.; Duan, Y.; Wang, W.; Donini, O.; Cieplak, P.; Srinivasan, J.; Case, D. A.; Cheatham, T. E. Calculating Structures and Free Energies of Complex Molecules: Combining Molecular Mechanics and Continuum Models. *Acc. Chem. Res.* **2000**, *33* (12), 889–897. https://doi.org/10.1021/ar000033j
95. Tanford, C.; Kirkwood, J. G. Theory of Protein Titration Curves. I. General Equations for Impenetrable Spheres. *J. Am. Chem. Soc.* **1957**, *79* (20), 5333–5339. https://doi.org/10.1021/ja01577a001
96. Warwicker, J.; Watson, H. C. Calculation of the Electric Potential in the Active Site Cleft Due to α-Helix Dipoles. *J. Mol. Biol.* **1982**, *157* (4), 671–679. https://doi.org/10.1016/0022-2836(82)90505-8
97. Gilson, M. K.; Honig, B. Calculation of the Total Electrostatic Energy of a Macromolecular System: Solvation Energies, Binding Energies, and Conformational Analysis. *Proteins: Struct., Funct., Bioinf.* **1988**, *4* (1), 7–18. https://doi.org/10.1002/prot.340040104
98. Gilson, M. K.; Sharp, K. A.; Honig, B. H. Calculating the Electrostatic Potential of Molecules in Solution: Method and Error Assessment.

J. Comput. Chem. **1988**, *9* (4), 327–335. https://doi.org/10.1002/jcc.540090407

99. Warshel, A.; Russell, S. T.; Churg, A. K. Macroscopic Models for Studies of Electrostatic Interactions in Proteins: Limitations and Applicability. *Proc. Natl. Acad. Sci. U.S.A.* **1984**, *81* (15), 4785–4789. https://doi.org/10.1073/pnas.81.15.4785
100. Hawkins, G. D.; Cramer, C. J.; Truhlar, D. G. Parametrized Models of Aqueous Free Energies of Solvation Based on Pairwise Descreening of Solute Atomic Charges from a Dielectric Medium. *J. Phys. Chem.* **1996**, *100* (51), 19824–19839. https://doi.org/10.1021/jp961710n
101. Qiu, D.; Shenkin, P. S.; Hollinger, F. P.; Still, W. C. The GB/SA Continuum Model for Solvation. A Fast Analytical Method for the Calculation of Approximate Born Radii. *J. Phys. Chem. A* **1997**, *101* (16), 3005–3014. https://doi.org/10.1021/jp961992r
102. Gallicchio, E.; Levy, R. M. AGBNP: An analytic implicit solvent model suitable for molecular dynamics simulations and high-resolution modeling. *J. Comput. Chem.* **2004**, *25* (4), 479–499. https://doi.org/10.1002/jcc.10400
103. Zhu, J.; Alexov, E.; Honig, B. Comparative Study of Generalized Born Models: Born Radii and Peptide Folding. *J. Phys. Chem. B* **2005**, *109* (7), 3008–3022. https://doi.org/10.1021/jp046307s
104. Ghosh, A.; Rapp, C. S.; Friesner, R. A. Generalized Born Model Based on a Surface Integral Formulation. *J. Phys. Chem. B* **1998**, *102* (52), 10983–10990. https://doi.org/10.1021/jp982533o
105. Fan, H.; Mark, A. E.; Zhu, J.; Honig, B. Comparative Study of Generalized Born Models: Protein Dynamics. *Proc. Natl. Acad. Sci. U.S.A.* **2005**, *102* (19), 6760–6764. https://doi.org/10.1073/pnas.0408857102
106. Onufriev, A.; Case, D. A.; Bashford, D. Effective Born Radii in the Generalized Born Approximation: The Importance of Being Perfect. *J. Comput. Chem.* **2002**, *23* (14), 1297–1304. https://doi.org/10.1002/jcc.10126

107. Borgis, D.; Lévy, N.; Marchi, M. Computing the Electrostatic Free-Energy of Complex Molecules: The Variational Coulomb Field Approximation. *J. Chem. Phys.* **2003**, *119* (6), 3516–3528. https://doi.org/10.1063/1.1589745
108. Orttung, W. H. Polarizability density of inert gas atom pairs. 1. *J. Chem.* **1985**, *89*, 3011–3016. https://doi.org/10.1021/j100260a012
109. Warshel, A. What about Protein Polarity? *Nature* **1987**, *330* (6143), 15–16. https://doi.org/10.1038/330015a0
110. Matthew, J. B.; Weber, P. C.; Salemme, F. R.; Richards, F. M. Electrostatic Orientation during Electron Transfer between Flavodoxin and Cytochrome c. *Nature* **1983**, *301* (5896), 169–171. https://doi.org/10.1038/301169a0
111. Bloomer, A. C.; Champness, J. N.; Bricogne, G.; Staden, R.; Klug, A. Protein Disk of Tobacco Mosaic Virus at 2.8 Å Resolution Showing the Interactions within and between Subunits. *Nature* **1978**, *276* (5686), 362–368. https://doi.org/10.1038/276362a0
112. Vernon, C. A.; Perutz, M. F. The Mechanisms of Hydrolysis of Glycosides and Their Relevance to Enzyme-Catalysed Reactions. *Proc. R. Soc. Lond. B* **1967**, *167* (1009), 389–401. https://doi.org/10.1098/rspb.1967.0036
113. Fothergill, M.; Goodman, M. F.; Petruska, J.; Warshel, A. Structure-Energy Analysis of the Role of Metal Ions in Phosphodiester Bond Hydrolysis by DNA Polymerase I. *J. Am. Chem. Soc.* **1995**, *117* (47), 11619–11627. https://doi.org/10.1021/ja00152a001
114. Madura, J. D.; Briggs, J. M.; Wade, R. C.; Davis, M. E.; Luty, B. A.; Ilin, A.; Antosiewicz, J.; Gilson, M. K.; Bagheri, B.; Scott, L. R.; McCammon, J. A. Electrostatics and Diffusion of Molecules in Solution: Simulations with the University of Houston Brownian Dynamics Program. *Comput. Phys. Commun.* **1995**, *91* (1), 57–95. https://doi.org/10.1016/0010-4655(95)00043-F
115. Baker, N. A.; Sept, D.; Joseph, S.; Holst, M. J.; McCammon, J. A. Electrostatics of Nanosystems: Application to Microtubules and the

Ribosome. *Proc. Natl. Acad. Sci. U.S.A.* **2001**, *98* (18), 10037–10041. https://doi.org/10.1073/pnas.181342398

116. Warshel, A.; Papazyan, A. Electrostatic Effects in Macromolecules: Fundamental Concepts and Practical Modeling. *Curr. Opin. Struct. Biol.* **1998**, *8* (2), 211–217. https://doi.org/10.1016/S0959-440X(98)80041-9
117. Alexov, E. G.; Gunner, M. R. Incorporating Protein Conformational Flexibility into the Calculation of pH-Dependent Protein Properties. *Biophys. J.* **1997**, *72* (5), 2075–2093. https://doi.org/10.1016/S0006-3495(97)78851-9
118. Alexov, E. G.; Gunner, M. R. Calculated Protein and Proton Motions Coupled to Electron Transfer: Electron Transfer from QA^- to QB in Bacterial Photosynthetic Reaction Centers. *Biochemistry* **1999**, *38* (26), 8253–8270. https://doi.org/10.1021/bi982700a
119. Ullmann, G. M.; Knapp, E.-W. Electrostatic Models for Computing Protonation and Redox Equilibria in Proteins. *Eur. Biophys. J.* **1999**, *28* (7), 533–551. https://doi.org/10.1007/s002490050236
120. Georgescu, R. E.; Alexov, E. G.; Gunner, M. R. Combining Conformational Flexibility and Continuum Electrostatics for Calculating pKas in Proteins. *Biophys. J.* **2002**, *83* (4), 1731–1748. https://doi.org/10.1016/S0006-3495(02)73940-4
121. Alden, R. G.; Parson, W. W.; Chu, Z. T.; Warshel, A. Calculations of Electrostatic Energies in Photosynthetic Reaction Centers. *J. Am. Chem. Soc.* **1995**, *117* (49), 12284–12298. https://doi.org/10.1021/ja00154a031
122. Johnson, E. T.; Parson, W. W. Electrostatic Interactions in an Integral Membrane Protein. *Biochemistry* **2002**, *41* (20), 6483–6494. https://doi.org/10.1021/bi012131y
123. Bashford, D.; Case, D. A. Generalized Born Models of Macromolecular Solvation Effects. *Annu. Rev. Phys. Chem.* **2000**, *51*, 129–152. https://doi.org/10.1146/annurev.physchem.51.1.129

124. Warshel, A.; Sharma, P. K.; Kato, M.; Parson, W. W. Modeling Electrostatic Effects in Proteins. *Biochim. Biophys. Acta Proteins Proteomics* **2006**, *1764* (11), 1647–1676. https://doi.org/10.1016/j.bbapap.2006.08.007
125. Warshel, A.; Russell, S. Theoretical Correlation of Structure and Energetics in the Catalytic Reaction of Trypsin. *J. Am. Chem. Soc.* **1986**, *108* (21), 6569–6579. https://doi.org/10.1021/ja00281a021
126. Marchi, M.; Gehlen, J. N.; Chandler, D.; Newton, M. Diabatic Surfaces and the Pathway for Primary Electron Transfer in a Photosynthetic Reaction Center. *J. Am. Chem. Soc.* **1993**, *115* (10), 4178–4190. https://doi.org/10.1021/ja00063a041
127. Stillinger, F. H.; Hasted, J. B.; Warshel, A.; Richards, R. E.; Franks, F. Theoretical Approaches to the Intermolecular Nature of Water. *Philos. Trans. R. Soc. Lond. B* **1977**, *278* (959), 97–112. https://doi.org/10.1098/rstb.1977.0033
128. Sharp, K. A.; Honig, B. Electrostatic Interactions in Macromolecules: Theory and Applications. *Annu. Rev. Biophys. Biomol. Struct.* **1990**, *19*, 301–332. https://doi.org/10.1146/annurev.bb.19.060190.001505
129. Warshel, A. Electrostatic Basis of Structure–Function Correlation in Proteins. *Acc. Chem. Res.* **1981**, *14* (9), 284–290. https://doi.org/10.1021/ar00069a004
130. Perutz, M. F. Electrostatic Effects in Proteins. *Science* **1978**, *201* (4362), 1187–1191. https://doi.org/10.1126/science.694508
131. Åqvist, J.; Fothergill, M. Computer Simulation of the Triosephosphate Isomerase Catalyzed Reaction (*). *J. Biol. Chem.* **1996**, *271* (17), 10010–10016. https://doi.org/10.1074/jbc.271.17.10010
132. Schutz, C. N.; Warshel, A. The Low Barrier Hydrogen Bond (LBHB) Proposal Revisited: The Case of the Asp···His Pair in Serine Proteases. *Proteins Struct., Funct., Bioinf.* **2004**, *55* (3), 711–723. https://doi.org/10.1002/prot.20096

133. Luzhkov, V. B.; Åqvist, J. A Computational Study of Ion Binding and Protonation States in the KcsA Potassium Channel. *Biochim. Biophys. Acta Protein Struct. Mol. Enzymol.* **2000**, *1481* (2), 360–370. https://doi.org/10.1016/S0167-4838(00)00183-7

134. Burykin, A.; Schutz, C. N.; Villà, J.; Warshel, A. Simulations of Ion Current in Realistic Models of Ion Channels: The KcsA Potassium Channel. *Proteins Struct., Funct., Bioinf.* **2002**, *47* (3), 265–280. https://doi.org/10.1002/prot.10106

135. Antosiewicz, J.; McCammon, J. A.; Gilson, M. K. Prediction of pH-Dependent Properties of Proteins. *J. Mol. Biol.* **1994**, *238* (3), 415–436. https://doi.org/10.1006/jmbi.1994.1301

136. Beroza, P.; Fredkin, D. R. Calculation of Amino Acid pKaS in a Protein from a Continuum Electrostatic Model: Method and Sensitivity Analysis. *J. Comput. Chem.* **1996**, *17* (10), 1229–1244. https://doi.org/10.1002/(SICI)1096-987X(19960730)17:10<1229::AID-JCC4>3.0.CO;2-Q

137. Koumanov, A.; Rüterjans, H.; Karshikoff, A. Continuum Electrostatic Analysis of Irregular Ionization and Proton Allocation in Proteins. *Proteins Struct., Funct., Bioinf.* **2002**, *46* (1), 85–96. https://doi.org/10.1002/prot.10034

138. Mehler, E. L.; Guarnieri, F. A Self-Consistent, Microenvironment Modulated Screened Coulomb Potential Approximation to Calculate pH-Dependent Electrostatic Effects in Proteins. *Biophys. J.* **1999**, *77* (1), 3–22. https://doi.org/10.1016/S0006-3495(99)76868-2

139. You, T. J.; Bashford, D. Conformation and Hydrogen Ion Titration of Proteins: A Continuum Electrostatic Model with Conformational Flexibility. *Biophys. J.* **1995**, *69* (5), 1721–1733. https://doi.org/10.1016/S0006-3495(95)80042-1

140. Zhou, H.-X.; Vijayakumar, M. Modeling of Protein Conformational Fluctuations in pKa Predictions. *J. Mol. Biol.* **1997**, *267* (4), 1002–1011. https://doi.org/10.1006/jmbi.1997.0895

141. van Vlijmen, H. W. T.; Schaefer, M.; Karplus, M. Improving the Accuracy of Protein pKa Calculations: Conformational Averaging versus the Average Structure. *Proteins Struct., Funct., Bioinf.* **1998**, *33* (2), 145–158. https://doi.org/10.1002/(SICI)1097-0134(19981101)33:2<145::AID-PROT1>3.0.CO;2-I
142. Dwyer, J. J.; Gittis, A. G.; Karp, D. A.; Lattman, E. E.; Spencer, D. S.; Stites, W. E.; García-Moreno, E. B. High Apparent Dielectric Constants in the Interior of a Protein Reflect Water Penetration. *Biophys. J.* **2000**, *79* (3), 1610–1620. https://doi.org/10.1016/S0006-3495(00)76411-3
143. Fitch, C. A.; Karp, D. A.; Lee, K. K.; Stites, W. E.; Lattman, E. E.; García-Moreno, E. B. Experimental pKa Values of Buried Residues: Analysis with Continuum Methods and Role of Water Penetration. *Biophys. J.* **2002**, *82* (6), 3289–3304. https://doi.org/10.1016/S0006-3495(02)75670-1
144. Kato, M.; Warshel, A. Using a Charging Coordinate in Studies of Ionization Induced Partial Unfolding. *J. Phys. Chem. B* **2006**, *110* (23), 11566–11570. https://doi.org/10.1021/jp061190o
145. Gao, J.; Kuczera, K.; Tidor, B.; Karplus, M. Hidden Thermodynamics of Mutant Proteins: A Molecular Dynamics Analysis. *Science* **1989**, *244* (4908), 1069–1072. https://doi.org/10.1126/science.2727695
146. Warshel, A.; Barboy, N. Energy Storage and Reaction Pathways in the First Step of the Vision Process. *J. Am. Chem. Soc.* **1982**, *104* (6), 1469–1476. https://doi.org/10.1021/ja00370a003
147. Perutz, M. F.; Kilmartin, J. V.; Nishikura, K.; Fogg, J. H.; Butler, P. J. G.; Rollema, H. S. Identification of Residues Contributing to the Bohr Effect of Human Haemoglobin. *J. Mol. Biol.* **1980**, *138* (3), 649–668. https://doi.org/10.1016/S0022-2836(80)80022-2
148. Warshel, A. Charge Stabilization Mechanism in the Visual and Purple Membrane Pigments. *Proc. Natl. Acad. Sci. U.S.A.* **1978**, *75* (6), 2558–2562. https://doi.org/10.1073/pnas.75.6.2558

149. Warshel, A. Calculations of Enzymic Reactions: Calculations of pKa, Proton Transfer Reactions, and General Acid Catalysis Reactions in Enzymes. *Biochemistry* **1981**, *20* (11), 3167–3177. https://doi.org/10.1021/bi00514a028
150. Jordan, F.; Li, H.; Brown, A. Remarkable Stabilization of Zwitterionic Intermediates May Account for a Billion-Fold Rate Acceleration by Thiamin Diphosphate-Dependent Decarboxylases. *Biochemistry* **1999**, *38* (20), 6369–6373. https://doi.org/10.1021/bi990373g
151. Dewar, M. J.; Storch, D. M. Alternative View of Enzyme Reactions. *Proc. Natl. Acad. Sci. U.S.A.* **1985**, *82* (8), 2225–2229. https://doi.org/10.1073/pnas.82.8.2225
152. Crosby, J.; Stone, R.; Lienhard, G. E. Mechanisms of Thiamine-Catalyzed Reactions. Decarboxylation of 2-(1-Carboxy-1-Hydroxyethyl)-3,4-Dimethylthiazolium Chloride. *J. Am. Chem. Soc.* **1970**, *92* (9), 2891–2900. https://doi.org/10.1021/ja00712a048
153. Lee, J. K.; Houk, K. N. A Proficient Enzyme Revisited: The Predicted Mechanism for Orotidine Monophosphate Decarboxylase. *Science* **1997**, *276* (5314), 942–945. https://doi.org/10.1126/science.276.5314.942
154. Warshel, A.; Štrajbl, M.; Villà, J.; Florián, J. Remarkable Rate Enhancement of Orotidine 5′-Monophosphate Decarboxylase Is Due to Transition-State Stabilization Rather than Ground-State Destabilization. *Biochemistry* **2000**, *39* (48), 14728–14738. https://doi.org/10.1021/bi000987h
155. Hwang, J.-K.; Warshel, A. Why Ion Pair Reversal by Protein Engineering Is Unlikely to Succeed. *Nature* **1988**, *334* (6179), 270–272. https://doi.org/10.1038/334270a0
156. Cutler, R. L.; Davies, A. M.; Creighton, S.; Warshel, A.; Moore, G. R.; Smith, M.; Mauk, A. G. Role of Arginine-38 in Regulation of the Cytochrome c Oxidation–Reduction Equilibrium. *Biochemistry* **1989**, *28* (8), 3188–3197. https://doi.org/10.1021/bi00434a012

157. Born, M. Volumen und Hydratationswärme der Ionen. *Z. Physik* **1920**, *1* (1), 45–48. https://doi.org/10.1007/BF01881023

158. Barril, X.; Alemán, C.; Orozco, M.; Luque, F. J. Salt Bridge Interactions: Stability of the Ionic and Neutral Complexes in the Gas Phase, in Solution, and in Proteins. *Proteins Struct., Funct., Bioinf.* **1998**, *32* (1), 67–79. https://doi.org/10.1002/(SICI)1097-0134(19980701)32:1<67::AID-PROT8>3.0.CO;2-B

159. Warshel, A.; Naray-Szabo, G. *Computational Approaches to Biochemical Reactivity.* Kluwer Academic Publishers, 1997.

160. Kumar, S.; Nussinov, R. Relationship between Ion Pair Geometries and Electrostatic Strengths in Proteins. *Biophys. J.* **2002**, *83* (3), 1595–1612. https://doi.org/10.1016/S0006-3495(02)73929-5

161. Hendsch, Z. S.; Tidor, B. Do Salt Bridges Stabilize Proteins? A Continuum Electrostatic Analysis. *Protein Sci.* **1994**, *3* (2), 211–226. https://doi.org/10.1002/pro.5560030206

162. Wada, A. The Alpha-Helix as an Electric Macro-Dipole. *Adv. Biophys.* **1976**, 1–63.

163. Hol, W. G. J.; van Duijnen, P. T.; Berendsen, H. J. C. The α-Helix Dipole and the Properties of Proteins. *Nature* **1978**, *273* (5662), 443–446. https://doi.org/10.1038/273443a0

164. Roux, B.; Bernèche, S.; Im, W. Ion Channels, Permeation, and Electrostatics: Insight into the Function of KcsA. *Biochemistry* **2000**, *39* (44), 13295–13306. https://doi.org/10.1021/bi001567v

165. Daggett, V. D.; Kollman, P. A.; Kuntz, I. D. Free Energy Perturbation Calculations of Charge Interactions with the Helix Dipole. *Chem. Scripta* **1989**, *29A*, 205–215.

166. van Duijnen, P. Th.; Thole, B. Th.; Hol, W. G. J. On the Role of the Active Site Helix in Papain, an Ab Initio Molecular Orbital Study. *Biophys. Chem.* **1979**, *9* (3), 273–280. https://doi.org/10.1016/0301-4622(79)85010-3

167. Gilson, M. K.; Honig, B. H. Energetics of Charge–Charge Interactions in Proteins. *Proteins Struct., Funct., Bioinf.* **1988**, *3* (1), 32–52. https://doi.org/10.1002/prot.340030104
168. Åqvist, J.; Luecke, H.; Quiocho, F. A.; Warshel, A. Dipoles Localized at Helix Termini of Proteins Stabilize Charges. *Proc. Natl. Acad. Sci. U.S.A.* **1991**, *88* (5), 2026–2030. https://doi.org/10.1073/pnas.88.5.2026
169. Lodi, P. J.; Knowles, J. R. Direct Evidence for the Exploitation of an α-Helix in the Catalytic Mechanism of Triosephosphate Isomerase. *Biochemistry* **1993**, *32*, 4338–4343. https://doi.org/10.1021/bi00067a024
170. Roux, B.; MacKinnon, R. The Cavity and Pore Helices in the KcsA K^+ Channel: Electrostatic Stabilization of Monovalent Cations. *Science* **1999**, *285* (5424), 100–102. https://doi.org/10.1126/science.285.5424.100
171. Burykin, A.; Kato, M.; Warshel, A. Exploring the Origin of the Ion Selectivity of the KcsA Potassium Channel. *Proteins Struct., Funct., Bioinf.* **2003**, *52* (3), 412–426. https://doi.org/10.1002/prot.10455
172. de Groot, B. L.; Grubmüller, H. The Dynamics and Energetics of Water Permeation and Proton Exclusion in Aquaporins. *Curr. Opin. Struct. Biol.* **2005**, *15* (2), 176–183. https://doi.org/10.1016/j.sbi.2005.02.003
173. Chakrabarti, N.; Roux, B.; Pomès, R. Structural Determinants of Proton Blockage in Aquaporins. *J. Mol. Biol.* **2004**, *343* (2), 493–510. https://doi.org/10.1016/j.jmb.2004.08.036
174. Kato, M.; Pisliakov, A. V.; Warshel, A. The Barrier for Proton Transport in Aquaporins as a Challenge for Electrostatic Models: The Role of Protein Relaxation in Mutational Calculations. *Proteins Struct., Funct., Bioinf.* **2006**, *64* (4), 829–844. https://doi.org/10.1002/prot.21012
175. Warshel, A.; Åqvist, J. Electrostatic Energy and Macromolecular Function. *Annu. Rev. Biophys. Biophys. Chem.* **1991**, *20*, 267–298. https://doi.org/10.1146/annurev.bb.20.060191.001411
176. King, G.; Lee, F. S.; Warshel, A. Microscopic Simulations of Macroscopic Dielectric Constants of Solvated Proteins. *J. Chem. Phys.* **1991**, *95* (6), 4366–4377. https://doi.org/10.1063/1.461760

177. Muegge, I.; Schweins, T.; Langen, R.; Warshel, A. Electrostatic Control of GTP and GDP Binding in the Oncoprotein P21ras. *Structure* **1996**, *4* (4), 475–489. https://doi.org/10.1016/S0969-2126(96)00052-4

178. Sham, Y. Y.; Muegge, I.; Warshel, A. The Effect of Protein Relaxation on Charge–Charge Interactions and Dielectric Constants of Proteins. *Biophys. J.* **1998**, *74* (4), 1744–1753. https://doi.org/10.1016/S0006-3495(98)77885-3

179. Nakamura, H. Roles of Electrostatic Interaction in Proteins. *Q. Rev. Biophys.* **1996**, *29* (1), 1–90. https://doi.org/10.1017/S0033583500005746

180. Simonson, T.; Brooks, C. L. Charge Screening and the Dielectric Constant of Proteins: Insights from Molecular Dynamics. *J. Am. Chem. Soc.* **1996**, *118* (35), 8452–8458. https://doi.org/10.1021/ja960884f

181. Smith, P. E.; Brunne, R. M.; Mark, A. E.; Van Gunsteren, W. F. Dielectric Properties of Trypsin Inhibitor and Lysozyme Calculated from Molecular Dynamics Simulations. *J. Phys. Chem.* **1993**, *97* (9), 2009–2014. https://doi.org/10.1021/j100111a046

182. Simonson, T.; Perahia, D.; Brünger, A. T. Microscopic Theory of the Dielectric Properties of Proteins. *Biophys. J.* **1991**, *59* (3), 670–690. https://doi.org/10.1016/S0006-3495(91)82282-2

183. Muegge, I.; Qi, P. X.; Wand, A. J.; Chu, Z. T.; Warshel, A. The Reorganization Energy of Cytochrome c Revisited. *J. Phys. Chem. B* **1997**, *101* (5), 825–836. https://doi.org/10.1021/jp962478o

184. Harvey, S. C.; Hoekstra, P. Dielectric Relaxation Spectra of Water Adsorbed on Lysozyme. *J. Phys. Chem.* **1972**, *76*, 2987–2994. https://doi.org/10.1021/j100665a011

185. Gilson, M. K.; Honig, B. H. The Dielectric Constant of a Folded Protein. *Biopolymers* **1986**, *25* (11), 2097–2119. https://doi.org/10.1002/bip.360251106

186. Cohen, B. E.; McAnaney, T. B.; Park, E. S.; Jan, Y. N.; Boxer, S. G.; Jan, L. Y. Probing Protein Electrostatics with a Synthetic Fluorescent Amino Acid. *Science* **2002**, *296* (5573), 1700–1703. https://doi.org/10.1126/science.1069346

187. Fluorescent Probe for Proteins. *Chem. Eng. News Arch.* **2002**, *80* (24), 30. https://doi.org/10.1021/cen-v080n024.p030
188. García-Moreno, E. B.; Dwyer, J. J.; Gittis, A. G.; Lattman, E. E.; Spencer, D. S.; Stites, W. E. Experimental Measurement of the Effective Dielectric in the Hydrophobic Core of a Protein. *Biophys. Chem.* **1997**, *64* (1), 211–224. https://doi.org/10.1016/S0301-4622(96)02238-7
189. Penfold, R.; Warwicker, J.; Jönsson, B. Electrostatic Models for Calcium Binding Proteins. *J. Phys. Chem. B* **1998**, *102* (43), 8599–8610. https://doi.org/10.1021/jp973420s
190. Mehler, E. L.; Eichele, G. Electrostatic Effects in Water-Accessible Regions of Proteins. *Biochemistry* **1984**, *23*, 3887–3891. https://doi.org/10.1021/bi00312a015
191. Debye, P.; Pauling, L. The Inter-Ionic Attraction Theory of Ionized Solutes. IV. The Influence of Variation of Dielectric Constant on the Limiting Law for Small Concentrations. *J. Am. Chem. Soc.* **1925**, *47* (8), 2129–2134. https://doi.org/10.1021/ja01685a008
192. Webb, T. J. The Free Energy of Hydration of Ions and the Electrostriction of the Solvent. *J. Am. Chem. Soc.* **1926**, *48* (10), 2589–2603. https://doi.org/10.1021/ja01421a013
193. Hill, T. L. The Electrostatic Contribution to Hindered Rotation in Certain Ions and Dipolar Ions in Solution. IV. *J. Chem. Phys.* **1944**, *12*, 147–156. https://doi.org/10.1063/1.1723927
194. Pearson, J. G.; Oldfield, E.; Lee, F. S.; Warshel, A. Chemical Shifts in Proteins: A Shielding Trajectory Analysis of the Fluorine NMR Spectrum of the E. coli Galactose-Binding Protein Using a Multipole Shielding Polarizability–Local Reaction Field–Molecular Dynamics Approach. *J. Am. Chem. Soc.* **1993**, *115* (15), 6851–6862. https://doi.org/10.1021/ja00068a049
195. Yang, A.-S.; Gunner, M. R.; Sampogna, R.; Sharp, K.; Honig, B. On the Calculation of pKa's in Proteins. *Proteins Struct., Funct., Bioinf.* **1993**, *15* (3), 252–265. https://doi.org/10.1002/prot.340150304

196. Warshel, A.; Sharma, P. K.; Kato, M.; Xiang, Y.; Liu, H.; Olsson, M. H. M. Electrostatic Basis for Enzyme Catalysis. *Chem. Rev.* **2006**, *106* (8), 3210–3235. https://doi.org/10.1021/cr0503106
197. Warshel, A. Computer Simulations of Enzyme Catalysis: Methods, Progress, and Insights. *Annu. Rev. Biophys. Biomol. Struct.* **2003**, *32*, 425–443. https://doi.org/10.1146/annurev.biophys.32.110601.141807
198. Fersht, A. *Structure and Mechanism in Protein Science. A Guide to Enzyme Catalysis and Protein Folding*; W. H. Freeman: New York, 1999.
199. Field, M. J. Simulating Enzyme Reactions: Challenges and Perspectives. *J. Comput. Chem.* **2002**, *23* (1), 48–58. https://doi.org/10.1002/jcc.1156
200. Alhambra, C.; Gao, J.; Corchado, J. C.; Villà, J.; Truhlar, D. G. Quantum Mechanical Dynamical Effects in an Enzyme-Catalyzed Proton Transfer Reaction. *J. Am. Chem. Soc.* **1999**, *121* (10), 2253–2258. https://doi.org/10.1021/ja9831655
201. Bash, P. A.; Field, M. J.; Davenport, R. C.; Petsko, G. A.; Ringe, D.; Karplus, M. Computer Simulation and Analysis of the Reaction Pathway of Triosephosphate Isomerase. *Biochemistry* **1991**, *30*, 5826–5832. https://doi.org/10.1021/bi00238a003
202. Hartsough, D. S.; Merz, K. M., Jr. Dynamic Force Field Models: Molecular Dynamics Simulations of Human Carbonic Anhydrase II Using a Quantum Mechanical/Molecular Mechanical Coupled Potential. *J. Phys. Chem.* **1995**, *99*, 11266–11275. https://doi.org/10.1021/j100028a030
203. Martí, S.; Andrés, J.; Moliner, V.; Silla, E.; Tuñón, I.; Bertrán, J. Transition Structure Selectivity in Enzyme Catalysis: A QM/MM Study of Chorismate Mutase. *Theor. Chem. Acc.* **2001**, *105* (3), 207–212. https://doi.org/10.1007/s002140000203
204. Mulholland, A. J.; Grant, G. H.; Richards, W. G. Computer Modelling of Enzyme Catalysed Reaction Mechanisms. *Protein Eng. Des. Sel.* **1993**, *6* (2), 133–147. https://doi.org/10.1093/protein/6.2.133

205. Singh, U. C.; Kollman, P. A. A Combined Ab Initio Quantum Mechanical and Molecular Mechanical Method for Carrying Out Simulations on Complex Molecular Systems: Applications to the $CH_3Cl + Cl^-$ Exchange Reaction and Gas Phase Protonation of Polyethers. *J. Comput. Chem.* **1986**, 7 (6), 718–730. https://doi.org/10.1002/jcc.540070604

206. Warshel, A. Energetics of Enzyme Catalysis. *Proc. Natl. Acad. Sci. U.S.A.* **1978**, *75* (11), 5250–5254. https://doi.org/10.1073/pnas.75.11.5250

207. Roca, M.; Martí, S.; Andrés, J.; Moliner, V.; Tuñón, I.; Bertrán, J.; Williams, I. H. Theoretical Modeling of Enzyme Catalytic Power: Analysis of "Cratic" and Electrostatic Factors in Catechol O-Methyltransferase. *J. Am. Chem. Soc.* **2003**, *125* (25), 7726–7737. https://doi.org/10.1021/ja0299497

208. Cannon, W. R.; Benkovic, S. J. Solvation, Reorganization Energy, and Biological Catalysis. *J. Biol. Chem.* **1998**, *273* (41), 26257–26260. https://doi.org/10.1074/jbc.273.41.26257

209. Náray-Szabó, G.; Warshel, A. *Computational Approaches to Biochemical Reactivity*; Springer, 2002.

210. Fisher, C. L.; Cabelli, D. E.; Hallewell, R. A.; Beroza, P.; Lo, T. P.; Getzoff, E. D.; Tainer, J. A. Computational, Pulse-Radiolytic, and Structural Investigations of Lys136 and Its Role in the Electrostatic Triad of Human Cu,Zn Superoxide Dismutase. *Proteins Struct., Funct., Bioinf.* **1997**, *29* (1), 103–112. https://doi.org/10.1002/(SICI)1097-0134(199709)29:1<103::AID-PROT8>3.0.CO;2-G

211. Antosiewicz, J.; Wlodek, S. T.; McCammon, J. A.; Gilson, M. K. Simulation of Charge-Mutant Acetylcholinesterases. *Biochemistry* **2000**, *39* (15), 4243–4250. https://doi.org/10.1021/bi000013a009

212. Fuxreiter, M.; Warshel, A. Origin of the Catalytic Power of Acetylcholinesterase: Computer Simulation Studies. *J. Am. Chem. Soc.* **1998**, *120* (1), 183–194. https://doi.org/10.1021/ja972326m

213. Cui, Q.; Elstner, M.; Kaxiras, E.; Frauenheim, T.; Karplus, M. A QM/MM Implementation of the Self-Consistent Charge Density Functional Tight Binding (SCC-DFTB) Method. *J. Phys. Chem. B* **2001**, *105* (2), 569–585. https://doi.org/10.1021/jp0029109

214. Fried, S. D.; Boxer, S. G. Electric Fields and Enzyme Catalysis. *Annu. Rev. Biochem.* **2017**, *86*, 387–415. https://doi.org/10.1146/annurev-biochem-061516-044432

215. Warshel, A.; Sharma, P. K.; Chu, Z. T.; Åqvist, J. Electrostatic Contributions to Binding of Transition-State Analogues Can Be Very Different from the Corresponding Contributions to Catalysis: Phenolates Binding to the Oxyanion Hole of Ketosteroid Isomerase. *Biochemistry* **2007**, *46* (6), 1466–1476. https://doi.org/10.1021/bi061752u

216. Kamerlin, S. C. L.; Sharma, P. K.; Chu, Z. T.; Warshel, A. Ketosteroid Isomerase Provides Further Support for the Idea That Enzymes Work by Electrostatic Preorganization. *Proc. Natl. Acad. Sci. U.S.A.* **2010**, *107* (9), 4075–4080. https://doi.org/10.1073/pnas.0914579107

217. Kraut, D. A.; Sigala, P. A.; Pybus, B.; Liu, C. W.; Ringe, D.; Petsko, G. A.; Herschlag, D. Testing Electrostatic Complementarity in Enzyme Catalysis: Hydrogen Bonding in the Ketosteroid Isomerase Oxyanion Hole. *PLoS Biol.* **2006**, *4* (4), e99. https://doi.org/10.1371/journal.pbio.0040099

218. Hammes-Schiffer, S. Quantum–Classical Simulation Methods for Hydrogen Transfer in Enzymes: A Case Study of Dihydrofolate Reductase. *Curr. Opin. Struct. Biol.* **2004**, *14* (2), 192–201. https://doi.org/10.1016/j.sbi.2004.03.008

219. Wong, K. F.; Selzer, T.; Benkovic, S. J.; Hammes-Schiffer, S. Impact of Distal Mutations on the Network of Coupled Motions Correlated to Hydride Transfer in Dihydrofolate Reductase. *Proc. Natl. Acad. Sci. U.S.A.* **2005**, *102* (19), 6807–6812. https://doi.org/10.1073/pnas.0408343102

220. Thorpe, I. F.; Brooks, C. L. Conformational Substates Modulate Hydride Transfer in Dihydrofolate Reductase. *J. Am. Chem. Soc.* **2005**, *127* (37), 12997–13006. https://doi.org/10.1021/ja053558l
221. Warshel, A.; Weiss, R. M. Energetics of Heme-Protein Interactions in Hemoglobin. *J. Am. Chem. Soc.* **1981**, *103* (2), 446–451. https://doi.org/10.1021/ja00392a033
222. Glennon, T. M.; Villà, J.; Warshel, A. How Does GAP Catalyze the GTPase Reaction of Ras? A Computer Simulation Study. *Biochemistry* **2000**, *39* (32), 9641–9651. https://doi.org/10.1021/bi000640e
223. Štrajbl, M.; Shurki, A.; Warshel, A. Converting Conformational Changes to Electrostatic Energy in Molecular Motors: The Energetics of ATP Synthase. *Proc. Natl. Acad. Sci. U.S.A.* **2003**, *100* (25), 14834–14839. https://doi.org/10.1073/pnas.2436328100
224. Shurki, A.; Warshel, A. Why Does the Ras Switch "Break" by Oncogenic Mutations? *Proteins Struct., Funct., Bioinf.* **2004**, *55* (1), 1–10. https://doi.org/10.1002/prot.20004
225. Xiang, Y.; Oelschlaeger, P.; Florián, J.; Goodman, M. F.; Warshel, A. Simulating the Effect of DNA Polymerase Mutations on Transition-State Energetics and Fidelity: Evaluating Amino Acid Group Contribution and Allosteric Coupling for Ionized Residues in Human Pol *β*. *Biochemistry* **2006**, *45* (23), 7036–7048. https://doi.org/10.1021/bi060147o
226. Jencks, W. P. *Catalysis in Chemistry and Enzymology*; Dover: New York, 1986.
227. van der Kamp, M. W.; Mulholland, A. J. Combined Quantum Mechanics/Molecular Mechanics (QM/MM) Methods in Computational Enzymology. *Biochemistry* **2013**, *52* (16), 2708–2728. https://doi.org/10.1021/bi400215w
228. Ferrer, S.; Moliner, V.; Świderek, K. Electrostatic Preorganization in Three Distinct Heterogeneous Proteasome *β*-Subunits. *ACS Catal.* **2024**, *14* (20), 15237–15249. https://doi.org/10.1021/acscatal.4c04964

229. Souza, B. S.; Mora, J. R.; Wanderlind, E. H.; Clementin, R. M.; Gesser, J. C.; Fiedler, H. D.; Nome, F.; Menger, F. M. Transforming a Stable Amide into a Highly Reactive One: Capturing the Essence of Enzymatic Catalysis. *Angew. Chem., Int. Ed.* **2017**, *56* (19), 5345–5348. https://doi.org/10.1002/anie.201701306

230. Warshel, A. Multiscale Modeling of Biological Functions: From Enzymes to Molecular Machines (Nobel Lecture). *Angew. Chem., Int. Ed.* **2014**, *53* (38), 10020–10031. https://doi.org/10.1002/anie.201403689

231. Jindal, G.; Warshel, A. Misunderstanding the Preorganization Concept Can Lead to Confusions about the Origin of Enzyme Catalysis. *Proteins Struct., Funct., Bioinf.* **2017**, *85* (12), 2157–2161. https://doi.org/10.1002/prot.25381

232. Johannin, G.; Kellershohn, N. An Estimate of Intraproteic Electrostatic Field Values Originated by the Peptide Groups in α-Chymotrypsin. *Biochem. Biophys. Res. Commun.* **1972**, *49* (2), 321–327. https://doi.org/10.1016/0006-291X(72)90413-5

233. Hayes, D. M.; Kollman, P. A. Electrostatic Potentials of Proteins. 1. Carboxypeptidase A. *J. Am. Chem. Soc.* **1976**, *98* (11), 3335–3345. https://doi.org/10.1021/ja00427a048

234. Vaissier, V.; Sharma, S. C.; Schaettle, K.; Zhang, T.; Head-Gordon, T. Computational Optimization of Electric Fields for Improving Catalysis of a Designed Kemp Eliminase. *ACS Catal.* **2018**, *8* (1), 219–227. https://doi.org/10.1021/acscatal.7b03151

235. Eberhart, M. E.; Wilson, T. R.; Jones, T. E.; Alexandrova, A. N. Electric Fields Imbue Enzyme Reactivity by Aligning Active-Site Fragment Orbitals. *Proc. Natl. Acad. Sci. U.S.A.* **2024**, *121* (44), e2411976121. https://doi.org/10.1073/pnas.2411976121

236. Roca, M.; Vardi-Kilshtain, A.; Warshel, A. Toward Accurate Screening in Computer-Aided Enzyme Design. *Biochemistry* **2009**, *48* (14), 3046–3056. https://doi.org/10.1021/bi802191b

237. Kassner, R. J. Effects of Nonpolar Environments on the Redox Potentials of Heme Complexes. *Proc. Natl. Acad. Sci. U.S.A.* **1972**, *69* (8), 2263–2267. https://doi.org/10.1073/pnas.69.8.2263
238. Zhou, H.-X. Control of Reduction Potential by Protein Matrix: Lesson from a Spherical Protein Model. *JBIC* **1997**, 2 (1), 109–113. https://doi.org/10.1007/s007750050113
239. Bertini, I.; Gori-Savellini, G.; Luchinat, C. Are Unit Charges Always Negligible? *JBIC* **1997**, 2 (1), 114–118. https://doi.org/10.1007/s007750050114
240. Mauk, A. G.; Moore, G. R. Control of Metalloprotein Redox Potentials: What Does Site-Directed Mutagenesis of Hemoproteins Tell Us? *JBIC* **1997**, 2 (1), 119–125. https://doi.org/10.1007/s007750050115
241. Gunner, M. R.; Alexov, E.; Torres, E.; Lipovaca, S. The Importance of the Protein in Controlling the Electrochemistry of Heme Metalloproteins: Methods of Calculation and Analysis. *JBIC* **1997**, 2 (1), 126–134. https://doi.org/10.1007/s007750050116
242. Náray-Szabó, G. Electrostatic Modulation of Electron Transfer in the Active Site of Heme Peroxidases. *JBIC* **1997**, 2 (1), 135–138. https://doi.org/10.1007/s007750050117
243. Armstrong, F. A. Evaluations of Reduction Potential Data in Relation to Coupling, Kinetics and Function. *JBIC* **1997**, 2 (1), 139–142. https://doi.org/10.1007/s007750050118
244. Rogers, N. K.; Moore, G. R. On the Energetics of Conformational Changes and pH-Dependent Redox Behavior of Electron Transfer Proteins. *FEBS Lett.* **1988**, *228* (1), 69–73. https://doi.org/10.1016/0014-5793(88)80587-8
245. Churg, A. K.; Weiss, R. M.; Warshel, A.; Takano, T. On the Action of Cytochrome c: Correlating Geometry Changes upon Oxidation with Activation Energies of Electron Transfer. *J. Phys. Chem.* **1983**, *87* (10), 1683–1694. https://doi.org/10.1021/j100233a010

246. Churg, A. K.; Warshel, A. Control of the Redox Potential of Cytochrome and Microscopic Dielectric Effects in Proteins. *Biochemistry* **1986**, *25*, 1675–1681. https://doi.org/10.1021/bi00355a035
247. Stephens, P. J.; Jollie, D. R.; Warshel, A. Protein Control of Redox Potentials of Iron–Sulfur Proteins. *Chem. Rev.* **1996**, *96* (7), 2491–2514. https://doi.org/10.1021/cr950045w
248. Swartz, P. D.; Beck, B. W.; Ichiye, T. Structural Origins of Redox Potentials in Fe–S Proteins: Electrostatic Potentials of Crystal Structures. *Biophys. J.* **1996**, *71* (6), 2958–2969. https://doi.org/10.1016/S0006-3495(96)79533-4
249. Rabenstein, B.; Ullmann, G. M.; Knapp, E. W. Energetics of Electron-Transfer and Protonation Reactions of the Quinones in the Photosynthetic Reaction Center of Rhodopseudomonas viridis. *Biochemistry* **1998**, *37* (8), 2488–2495. https://doi.org/10.1021/bi971921y
250. Noodleman, L.; Lovell, T.; Liu, T.; Himo, F.; Torres, R. A. Insights into Properties and Energetics of Iron–Sulfur Proteins from Simple Clusters to Nitrogenase. *Curr. Opin. Chem. Biol.* **2002**, *6* (2), 259–273. https://doi.org/10.1016/S1367-5931(02)00309-5
251. Teixeira, V. H.; Soares, C. M.; Baptista, A. M. Studies of the Reduction and Protonation Behavior of Tetraheme Cytochromes Using Atomic Detail. *J. Biol. Inorg. Chem.* **2002**, *7* (1), 200–216. https://doi.org/10.1007/s007750100287
252. Kanda, T.; Ishikita, H. Energetics of the Electron Transfer Pathways in the Homodimeric Photosynthetic Reaction Center. *Biochemistry* **2022**, *61* (22), 2621–2627. https://doi.org/10.1021/acs.biochem.2c00524
253. Yelle, R. B.; Ichiye, T. Solvation Free-Energy Reaction Curves for Electron Transfer in Aqueous Solution: Theory and Simulation. *J. Phys. Chem. B* **1997**, *101* (20), 4127–4135. https://doi.org/10.1021/jp970081u
254. Blumberger, J. Recent Advances in the Theory and Molecular Simulation of Biological Electron Transfer Reactions. *Chem. Rev.* **2015**, *115* (20), 11191–11238. https://doi.org/10.1021/acs.chemrev.5b00298

255. Muegge, I.; Qi, P. X.; Wand, A. J.; Chu, Z. T.; Warshel, A. The Reorganization Energy of Cytochrome c Revisited. *J. Phys. Chem. B* **1997**, *101* (5), 825–836. https://doi.org/10.1021/jp962478o
256. Deisenhofer, J.; Epp, O.; Miki, K.; Huber, R.; Michel, H. Structure of the Protein Subunits in the Photosynthetic Reaction Centre of Rhodopseudomonas viridis at 3 Å Resolution. *Nature* **1985**, *318* (6047), 618–624. https://doi.org/10.1038/318618a0
257. Deisenhofer, J.; Epp, O.; Sinning, I.; Michel, H. Crystallographic Refinement at 2.3 Å Resolution and Refined Model of the Photosynthetic Reaction Centre from Rhodopseudomonas viridis. *J. Mol. Biol.* **1995**, *246* (3), 429–457. https://doi.org/10.1006/jmbi.1994.0097
258. Chang, C. H.; El-Kabbani, O.; Tiede, D.; Norris, J.; Schiffer, M. Structure of the Membrane-Bound Protein Photosynthetic Reaction Center from Rhodobacter sphaeroides. *Biochemistry* **2000**, *39* (6), 1300–1310. https://doi.org/10.1021/bi00236a005
259. Allen, J. P.; Feher, G.; Yeates, T. O.; Komiya, H.; Rees, D. C. Structure of the Reaction Center from Rhodobacter sphaeroides R-26: The Cofactors. *Proc. Natl. Acad. Sci. U.S.A.* **1987**, *84* (16), 5730–5734. https://doi.org/10.1073/pnas.84.16.5730
260. Treutlein, H.; Schulten, K.; Deisenhofer, J.; Michel, H.; Brünger, A.; Karplus, M. Molecular Dynamics Simulation of the Primary Processes in the Photosynthetic Reaction Center of Rhodopseudomonas viridis. In *The Photosynthetic Bacterial Reaction Center: Structure and Dynamics*; Breton, J.; Verméglio, A., Eds.; Springer US: Boston, MA, 1988; pp 139–150. https://doi.org/10.1007/978-1-4899-0815-5_16
261. Treutlein, H.; Schulten, K.; Niedermeier, C.; Deisenhofer, J.; Michel, H.; DeVault, D. Electrostatic Control of Electron Transfer in the Photosynthetic Reaction Center of Rhodopseudomonas viridis. In *The Photosynthetic Bacterial Reaction Center: Structure and Dynamics*; Breton, J.; Verméglio, A., Eds.; Springer US: Boston, MA, 1988; pp 369–377. https://doi.org/10.1007/978-1-4899-0815-5_37

262. Creighton, S.; Hwang, J. K.; Warshel, A.; Parson, W. W.; Norris, J. Simulating the Dynamics of the Primary Charge Separation Process in Bacterial Photosynthesis. *Biochemistry* **1988**, *27* (2), 774–781. https://doi.org/10.1021/bi00402a043

263. Parson, W. W.; Chu, Z.-T.; Warshel, A. Electrostatic Control of Charge Separation in Bacterial Photosynthesis. *Biochim. Biophys. Acta Bioenerg.* **1990**, *1017* (3), 251–272. https://doi.org/10.1016/0005-2728(90)90192-7

264. Alden, R. G.; Parson, W. W.; Chu, Z. T.; Warshel, A. Orientation of the OH Dipole of Tyrosine (M)210 and Its Effect on Electrostatic Energies in Photosynthetic Bacterial Reaction Centers. *J. Phys. Chem.* **1996**, *100* (41), 16761–16770. https://doi.org/10.1021/jp961271s

265. Chuang, J. I.; Boxer, S. G.; Holten, D.; Kirmaier, C. High Yield of M-Side Electron Transfer in Mutants of Rhodobacter capsulatus Reaction Centers Lacking the L-Side Bacteriopheophytin. *Biochemistry* **2006**, *45* (12), 3845–3851. https://doi.org/10.1021/bi0601048

266. Yarnell, A. Photosynthesis Mystery Cracked. *Chem. Eng. News* **2006**, *84* (13), 30.

267. Warshel, A.; Churg, A. K.; Huber, R. Converting Structural Changes upon Oxidation of Cytochrome c to Electrostatic Reorganization Energy. *J. Mol. Biol.* **1983**, *168* (3), 693–697. https://doi.org/10.1016/S0022-2836(83)80310-6

268. Muegge, I.; Rarey, M. Small Molecule Docking and Scoring. In *Rev. Comput. Chem.*; John Wiley & Sons, Ltd.: New York, 2001; Vol. 18, pp 1–60. https://doi.org/10.1002/0471224413.ch1

269. Kollman, P. A.; Massova, I.; Reyes, C.; Kuhn, B.; Huo, S.; Chong, L.; Lee, M.; Lee, T.; Duan, Y.; Wang, W.; Donini, O.; Cieplak, P.; Srinivasan, J.; Case, D. A.; Cheatham, T. E. Calculating Structures and Free Energies of Complex Molecules: Combining Molecular Mechanics and Continuum Models. *Acc. Chem. Res.* **2000**, *33* (12), 889–897. https://doi.org/10.1021/ar000033j

270. Mondal, D.; Florián, J.; Warshel, A. Exploring the Effectiveness of Binding Free Energy Calculations. *J. Phys. Chem. B* **2019**, *123* (42), 8910–8915. https://doi.org/10.1021/acs.jpcb.9b07593

271. Wang, L.; Wu, Y.; Deng, Y.; Kim, B.; Pierce, L.; et al. Accurate and Reliable Prediction of Relative Ligand Binding Potency in Prospective Drug Discovery by Way of a Modern Free-Energy Calculation Protocol and Force Field. *J. Am. Chem. Soc.* **2015**, *137* (7), 2695–2703. https://doi.org/10.1021/ja512751q

272. Lee, F. S.; Chu, Z.-T.; Bolger, M. B.; Warshel, A. Calculations of Antibody–Antigen Interactions: Microscopic and Semi-Microscopic Evaluation of the Free Energies of Binding of Phosphorylcholine Analogs to McPC603. *Protein Eng. Des. Sel.* **1992**, *5* (3), 215–228. https://doi.org/10.1093/protein/5.3.215

273. Sham, Y. Y.; Chu, Z. T.; Tao, H.; Warshel, A. Examining Methods for Calculations of Binding Free Energies: LRA, LIE, PDLD-LRA, and PDLD/S-LRA Calculations of Ligands Binding to an HIV Protease. *Proteins Struct., Funct., Bioinf.* **2000**, *39* (4), 393–407. https://doi.org/10.1002/(SICI)1097-0134(20000601)39:4<393::AID-PROT120>3.0.CO;2-H

274. Åqvist, J.; Hansson, T. On the Validity of Electrostatic Linear Response in Polar Solvents. *J. Phys. Chem.* **1996**, *100* (22), 9512–95221. https://doi.org/10.1021/jp953640a

275. Singh, N.; Warshel, A. Absolute Binding Free Energy Calculations: On the Accuracy of Computational Scoring of Protein–Ligand Interactions. *Proteins Struct., Funct., Bioinf.* **2010**, *78* (7), 1705–1723. https://doi.org/10.1002/prot.22687

276. Still, W. C.; Tempczyk, A.; Hawley, R. C.; Hendrickson, T. Semianalytical Treatment of Solvation for Molecular Mechanics and Dynamics. *J. Am. Chem. Soc.* **1990**, *112* (16), 6127–6129. https://doi.org/10.1021/ja00172a038

277. Florián, J.; Goodman, M. F.; Warshel, A. Theoretical Investigation of the Binding Free Energies and Key Substrate-Recognition Components of the Replication Fidelity of Human DNA Polymerase *β*. *J. Phys. Chem. B* **2002**, *106* (22), 5739–5753. https://doi.org/10.1021/jp020790u
278. Massova, I.; Kollman, P. A. Combined Molecular Mechanical and Continuum Solvent Approach (MM-PBSA/GBSA) to Predict Ligand Binding. *Perspect. Drug Discov. Des.* **2000**, *18* (1), 113–135. https://doi.org/10.1023/A:1008763014207
279. Bakowies, D.; van Gunsteren, W. F. Simulations of Apo and Holo–Fatty Acid Binding Protein: Structure and Dynamics of Protein, Ligand, and Internal Water. *J. Mol. Biol.* **2002**, *315* (4), 713–736. https://doi.org/10.1006/jmbi.2001.5202
280. Karplus, M.; Kushick, J. N. Method for Estimating the Configurational Entropy of Macromolecules. *Macromolecules* **1981**, *14* (2), 325–332. https://doi.org/10.1021/ma50003a019
281. Warshel, A. Office of Naval Research, Award Number N00014-91-J-1318. (Not published)
282. Hermans, J.; Wang, L. Inclusion of Loss of Translational and Rotational Freedom in Theoretical Estimates of Free Energies of Binding. Application to a Complex of Benzene and Mutant T4 Lysozyme. *J. Am. Chem. Soc.* **1997**, *119* (11), 2707–2714. https://doi.org/10.1021/ja963568+
283. Štrajbl, M.; Sham, Y. Y.; Villà, J.; Chu, Z.-T.; Warshel, A. Calculations of Activation Entropies of Chemical Reactions in Solution. *J. Phys. Chem. B* **2000**, *104* (18), 4578–4584. https://doi.org/10.1021/jp0003095
284. Singh, N.; Warshel, A. A Comprehensive Examination of the Contributions to the Binding Entropy of Protein–Ligand Complexes. *Proteins Struct., Funct., Bioinf.* **2010**, *78* (7), 1724–1735. https://doi.org/10.1002/prot.22689
285. Villà, J.; Štrajbl, M.; Glennon, T. M.; Sham, Y. Y.; Chu, Z.-T.; Warshel, A. How Important Are Entropic Contributions to Enzyme Catalysis?

Proc. Natl. Acad. Sci. U.S.A. **2000**, *97* (22), 11899–11904. https://doi.org/10.1073/pnas.97.22.11899

286. Sharma, P. K.; Xiang, Y.; Kato, M.; Warshel, A. What Are the Roles of Substrate-Assisted Catalysis and Proximity Effects in Peptide Bond Formation by the Ribosome? *Biochemistry* **2005**, *44* (34), 11307–11314. https://doi.org/10.1021/bi0509806
287. Chakrabarty, S.; Warshel, A. Capturing the Energetics of Water Insertion in Biological Systems: The Water Flooding Approach. *Proteins Struct., Funct., Bioinf.* **2013**, *81* (1), 93–106. https://doi.org/10.1002/prot.24165
288. Yoon, H.; Kolev, V.; Warshel, A. Validating the Water Flooding Approach by Comparing It to Grand Canonical Monte Carlo Simulations. *J. Phys. Chem. B* **2017**, *121* (40), 9358–9365. https://doi.org/10.1021/acs.jpcb.7b07726
289. Kornberg, A.; Baker, T. A. *DNA Replication*; W. H. Freeman: New York, 1992.
290. Schroeder, G. K.; Lad, C.; Wyman, P.; Williams, N. H.; Wolfenden, R. The Time Required for Water Attack at the Phosphorus Atom of Simple Phosphodiesters and of DNA. *Proc. Natl. Acad. Sci. U.S.A.* **2006**, *103* (11), 4052–4055. https://doi.org/10.1073/pnas.0510879103
291. Wolfenden, R.; Snider, M. J. The Depth of Chemical Time and the Power of Enzymes as Catalysts. *Acc. Chem. Res.* **2001**, *34* (12), 938–945. https://doi.org/10.1021/ar000058i
292. Echols, H.; Goodman, M. F. Fidelity Mechanisms in DNA Replication. *Annu. Rev. Biochem.* **1991**, *60*, 477–511.
293. Johnson, K. Conformational Coupling in DNA Polymerase Fidelity. *Annu. Rev. Biochem.* **1993**, *62*, 685–713. https://doi.org/10.1146/annurev.bi.62.070193.003345
294. Joyce, C. M.; Benkovic, S. J. DNA Polymerase Fidelity: Kinetics, Structure, and Checkpoints. *Biochemistry* **2004**, *43* (45), 14317–14324. https://doi.org/10.1021/bi048422z

295. Steitz, T. A.; Smerdon, S. J.; Jäger, J.; Joyce, C. M. A Unified Polymerase Mechanism for Nonhomologous DNA and RNA Polymerases. *Science* **1994**, *266* (5193), 2022–2025. https://doi.org/10.1126/science.7528445
296. Florián, J.; Goodman, M. F.; Warshel, A. Computer Simulation of the Chemical Catalysis of DNA Polymerases: Discriminating between Alternative Nucleotide Insertion Mechanisms for T7 DNA Polymerase. *J. Am. Chem. Soc.* **2003**, *125* (27), 8163–8177. https://doi.org/10.1021/ja028997o
297. Florián, J.; Goodman, M. F.; Warshel, A. Computer Simulation Studies of the Fidelity of DNA Polymerases. *Biopolymers* **2003**, *68* (3), 286–299. https://doi.org/10.1002/bip.10244
298. Ram Prasad, B.; Warshel, A. Prechemistry versus Preorganization in DNA Replication Fidelity. *Proteins Struct., Funct., Bioinf.* **2011**, *79* (10), 2900–2919. https://doi.org/10.1002/prot.23128
299. Beard, W. A.; Wilson, S. H. Structural Insights into the Origins of DNA Polymerase Fidelity. *Structure* **2003**, *11* (5), 489–496. https://doi.org/10.1016/S0969-2126(03)00051-0
300. Goodman, M. F.; Creighton, S.; Bloom, L. B.; Petruska, J.; Kunkel, T. A. Biochemical Basis of DNA Replication Fidelity. *Crit. Rev. Biochem. Mol. Biol.* **1993**, *28* (2), 83–126. https://doi.org/10.3109/10409239309086792
301. Showalter, A. K.; Lamarche, B. J.; Bakhtina, M.; Su, M.-I.; Tang, K.-H.; Tsai, M.-D. Mechanistic Comparison of High-Fidelity and Error-Prone DNA Polymerases and Ligases Involved in DNA Repair. *Chem. Rev.* **2006**, *106* (2), 340–360. https://doi.org/10.1021/cr040487k
302. Xiang, Y.; Oelschlaeger, P.; Florián, J.; Goodman, M. F.; Warshel, A. Simulating the Effect of DNA Polymerase Mutations on Transition-State Energetics and Fidelity: Evaluating Amino Acid Group Contribution and Allosteric Coupling for Ionized Residues in Human Pol *β*. *Biochemistry* **2006**, *45* (23), 7036–7048. https://doi.org/10.1021/bi060147o
303. Kamerlin, S. C. L.; Sharma, P. K.; Prasad, R. B.; Warshel, A. Why Nature Really Chose Phosphate. *Quart. Rev. Biophys.* **2013**, *46* (1), 1–132. https://doi.org/10.1017/S0033583512000157

304. Yang, L.; Beard, W. A.; Wilson, S. H.; Broyde, S.; Schlick, T. Highly Organized but Pliant Active Site of DNA Polymerase β: Compensatory Mechanisms in Mutant Enzymes Revealed by Dynamics Simulations and Energy Analyses. *Biophys. J.* **2004**, *86* (6), 3392–3408. https://doi.org/10.1529/biophysj.103.036012
305. Feig, M.; Zacharias, M.; Pettitt, B. M. Conformations of an Adenine Bulge in a DNA Octamer and Its Influence on DNA Structure from Molecular Dynamics Simulations. *Biophys. J.* **2001**, *81* (1), 352–370. https://doi.org/10.1016/S0006-3495(01)75705-0
306. Oostenbrink, C.; van Gunsteren, W. F. Efficient Calculation of Many Stacking and Pairing Free Energies in DNA from a Few Molecular Dynamics Simulations. *Chem. — Eur. J.* **2005**, *11* (15), 4340–4348. https://doi.org/10.1002/chem.200401120
307. Orozco, M.; Pérez, A.; Noy, A.; Luque, F. J. Theoretical Methods for the Simulation of Nucleic Acids. *Chem. Soc. Rev.* **2003**, *32* (6), 350–364. https://doi.org/10.1039/B207226M
308. Seibert, E.; Ross, J. B. A.; Osman, R. Contribution of Opening and Bending Dynamics to Specific Recognition of DNA Damage. *J. Mol. Biol.* **2003**, *330* (4), 687–703. https://doi.org/10.1016/S0022-2836(03)00598-9
309. Arora, K.; Beard, W. A.; Wilson, S. H.; Schlick, T. Mismatch-Induced Conformational Distortions in Polymerase β Support an Induced-Fit Mechanism for Fidelity. *Biochemistry* **2005**, *44* (40), 13328–13341. https://doi.org/10.1021/bi0507682
310. Radhakrishnan, R.; Schlick, T. Orchestration of Cooperative Events in DNA Synthesis and Repair Mechanism Unraveled by Transition Path Sampling of DNA Polymerase β's Closing. *Proc. Natl. Acad. Sci. U.S.A.* **2004**, *101* (16), 5970–5975. https://doi.org/10.1073/pnas.0308585101
311. Radhakrishnan, R.; Schlick, T. Fidelity Discrimination in DNA Polymerase β: Differing Closing Profiles for a Mismatched (G:A) versus Matched (G:C) Base Pair. *J. Am. Chem. Soc.* **2005**, *127* (38), 13245–13252. https://doi.org/10.1021/ja052623o

312. Alberts, I. L.; Wang, Y.; Schlick, T. DNA Polymerase β Catalysis: Are Different Mechanisms Possible? *J. Am. Chem. Soc.* **2007**, *129* (36), 11100–11110. https://doi.org/10.1021/ja071533b

313. Florián, J.; Goodman, M. F.; Warshel, A. Computer Simulations of Protein Functions: Searching for the Molecular Origin of the Replication Fidelity of DNA Polymerases. *Proc. Natl. Acad. Sci. U.S.A.* **2005**, *102* (19), 6819–6824. https://doi.org/10.1073/pnas.0408173102

314. Xiang, Y.; Goodman, M. F.; Beard, W. A.; Wilson, S. H.; Warshel, A. Exploring the Role of Large Conformational Changes in the Fidelity of DNA Polymerase β. *Proteins* **2008**, *70* (1), 231–247. https://doi.org/10.1002/prot.21668

315. Kamerlin, S. C. L.; Warshel, A. At the Dawn of the 21st Century: Is Dynamics the Missing Link for Understanding Enzyme Catalysis? *Proteins* **2010**, *78* (6), 1339–1375. https://doi.org/10.1002/prot.22654

316. Lin, P.; Pedersen, L. C.; Batra, V. K.; Beard, W. A.; Wilson, S. H.; Pedersen, L. G. Energy Analysis of Chemistry for Correct Insertion by DNA Polymerase β. *Proc. Natl. Acad. Sci. U.S.A.* **2006**, *103* (36), 13294–13299. https://doi.org/10.1073/pnas.0606006103

317. Lin, P.; Batra, V. K.; Pedersen, L. C.; Beard, W. A.; Wilson, S. H.; Pedersen, L. G. Incorrect Nucleotide Insertion at the Active Site of a G:A Mismatch Catalyzed by DNA Polymerase β. *Proc. Natl. Acad. Sci. U.S.A.* **2007**, *105* (15), 5670–5674. https://doi.org/10.1073/pnas.0801257105

318. Rucker, R.; Oelschlaeger, P.; Warshel, A. A Binding Free Energy Decomposition Approach for Accurate Calculations of the Fidelity of DNA Polymerases. *Proteins* **2009**, *78* (3), 671–680. https://doi.org/10.1002/prot.22596

319. Radhakrishnan, R.; Arora, K.; Wang, Y.; Beard, W. A.; Wilson, S. H.; Schlick, T. Regulation of DNA Repair Fidelity by Molecular Checkpoints: "Gates" in DNA Polymerase β's Substrate Selection. *Biochemistry* **2006**, *45* (51), 15142–15156. https://doi.org/10.1021/bi061353z

320. Radhakrishnan, R.; Schlick, T. Correct and Incorrect Nucleotide Incorporation Pathways in DNA Polymerase *β*. *Biochem. Biophys. Res. Commun.* **2006**, *350* (3), 521–529. https://doi.org/10.1016/j.bbrc.2006.09.059
321. Wang, Y.; Schlick, T. Quantum Mechanics/Molecular Mechanics Investigation of the Chemical Reaction in Dpo4 Reveals Water-Dependent Pathways and Requirements for Active Site Reorganization. *J. Am. Chem. Soc.* **2008**, *130* (40), 13240–13250. https://doi.org/10.1021/ja802215c
322. Klähn, M.; Braun-Sand, S.; Rosta, E.; Warshel, A. On Possible Pitfalls in Ab Initio Quantum Mechanics/Molecular Mechanics Minimization Approaches for Studies of Enzymatic Reactions. *J. Phys. Chem. B* **2005**, *109* (32), 15645–15650. https://doi.org/10.1021/jp0521757
323. Bakhtina, M.; Lee, S.; Wang, Y.; Dunlap, C.; Lamarche, B.; Tsai, M.-D. Use of Viscogens, dNTPαS, and Rhodium(III) as Probes in Stopped-Flow Experiments To Obtain New Evidence for the Mechanism of Catalysis by DNA Polymerase *β*. *Biochemistry* **2005**, *44* (13), 5177–5187. https://doi.org/10.1021/bi047664w
324. Abashkin, Y. G.; Erickson, J. W.; Burt, S. K. Quantum Chemical Investigation of Enzymatic Activity in DNA Polymerase *β*. A Mechanistic Study. *J. Phys. Chem. B* **2001**, *105* (1), 287–292. https://doi.org/10.1021/jp003629x
325. Bojin, M. D.; Schlick, T. A Quantum Mechanical Investigation of Possible Mechanisms for the Nucleotidyl Transfer Reaction Catalyzed by DNA Polymerase *β*. *J. Phys. Chem. B* **2007**, *111* (38), 11244–11252. https://doi.org/10.1021/jp071838c
326. Rittenhouse, R. C.; Apostoluk, W. K.; Miller, J. H.; Straatsma, T. P. Characterization of the Active Site of DNA Polymerase *β* by Molecular Dynamics and Quantum Chemical Calculation. *Proteins* **2003**, *53* (3), 667–682. https://doi.org/10.1002/prot.10451
327. Lee, F. S.; Chu, Z. T.; Warshel, A. Microscopic and Semimicroscopic Calculations of Electrostatic Energies in Proteins by the POLARIS and ENZYMIX Programs. *J. Comput. Chem.* **1993**, *14* (2), 161–185. https://doi.org/10.1002/jcc.540140205

328. Arora, K.; Schlick, T. In Silico Evidence for DNA Polymerase-β's Substrate-Induced Conformational Change. *Biophys. J.* **2004**, *87* (5), 3088–3099. https://doi.org/10.1529/biophysj.104.040915

329. Sampoli Benítez, B. A.; Arora, K.; Schlick, T. In Silico Studies of the African Swine Fever Virus DNA Polymerase X Support an Induced-Fit Mechanism. *Biophys. J.* **2006**, *90* (1), 42–56. https://doi.org/10.1529/biophysj.105.071944

330. Koshland, D. E. Jr. The Key–Lock Theory and the Induced Fit Theory. *Angew. Chem. Int. Ed. Engl.* **1995**, *33* (23–24), 2375–2378. https://doi.org/10.1002/anie.199423751

331. Post, C. B.; Ray, W. J. Jr. Reexamination of Induced Fit as a Determinant of Substrate Specificity in Enzymic Reactions. *Biochemistry* **1995**, *34* (49), 15881–15885. https://doi.org/10.1021/bi00049a001

332. Sawaya, M. R.; Prasad, R.; Wilson, S. H.; Kraut, J.; Pelletier, H. Crystal Structures of Human DNA Polymerase β Complexed with Gapped and Nicked DNA: Evidence for an Induced Fit Mechanism. *Biochemistry* **1997**, *36* (37), 11205–11215. https://doi.org/10.1021/bi9703812

333. Warshel, A.; Sharma, P. K.; Kato, M.; Xiang, Y.; Liu, H.; Olsson, M. H. M. Electrostatic Basis for Enzyme Catalysis. *Chem. Rev.* **2006**, *106* (8), 3210–3235. https://doi.org/10.1021/cr0503106

334. Liu, H.; Warshel, A. The Catalytic Effect of Dihydrofolate Reductase and Its Mutants Is Determined by Reorganization Energies. *Biochemistry* **2007**, *46* (20), 6011–6025. https://doi.org/10.1021/bi700201w

335. Lamm, G.; Pack, G. R. Counterion Condensation and Shape within Poisson–Boltzmann Theory. *Biopolymers* **2010**, *93* (7), 619–639. https://doi.org/10.1002/bip.21421

336. Jaenicke, R. Protein Stability and Molecular Adaptation to Extreme Conditions. *Eur. J. Biochem.* **1991**, *202* (3), 715–728. https://doi.org/10.1111/j.1432-1033.1991.tb16426.x

337. Smith, L. J.; Jones, R. M.; van Gunsteren, W. F. Characterization of the Denaturation of Human α-Lactalbumin in Urea by Molecular Dynamics

Simulations. *Proteins* **2005**, *58* (2), 439–449. https://doi.org/10.1002/prot.20287

338. Go, N. Theoretical Studies of Protein Folding. *Annu. Rev. Biophys.* **1983**, *12*, 183–210. https://doi.org/10.1146/annurev.bb.12.060183.001151
339. Onuchic, J. N.; Wolynes, P. G.; Luthey-Schulten, Z.; Socci, N. D. Toward an Outline of the Topography of a Realistic Protein-Folding Funnel. *Proc. Natl. Acad. Sci. U.S.A.* **1995**, *92* (8), 3626–3630. https://doi.org/10.1073/pnas.92.8.3626
340. Karanicolas, J.; Brooks, C. L. The Structural Basis for Biphasic Kinetics in the Folding of the WW Domain from a Formin-Binding Protein: Lessons for Protein Design? *Proc. Natl. Acad. Sci. U.S.A.* **2003**, *100* (7), 3954–3959. https://doi.org/10.1073/pnas.0731771100
341. Dill, K. A. Dominant Forces in Protein Folding. *Biochemistry* **1990**, *29* (31), 7133–7155. https://doi.org/10.1021/bi00483a001
342. Snow, C. D.; Nguyen, H.; Pande, V. S.; Gruebele, M. Absolute Comparison of Simulated and Experimental Protein-Folding Dynamics. *Nature* **2002**, *420* (6911), 102–106. https://doi.org/10.1038/nature01160
343. Fan, Z. Z.; Hwang, J.-K.; Warshel, A. Using Simplified Protein Representation as a Reference Potential for All-Atom Calculations of Folding Free Energy. *Theor. Chem. Acc.* **1999**, *103*, 77–80. https://doi.org/10.1007/s002140050516
344. Khalili, M.; Liwo, A.; Scheraga, H. A. Kinetic Studies of Folding of the B-Domain of Staphylococcal Protein A with Molecular Dynamics and a United-Residue (UNRES) Model of Polypeptide Chains. *J. Mol. Biol.* **2006**, *355* (3), 536–547. https://doi.org/10.1016/j.jmb.2005.10.056
345. Hollien, J.; Marqusee, S. Structural Distribution of Stability in a Thermophilic Enzyme. *Proc. Natl. Acad. Sci. U.S.A.* **1999**, *96* (24), 13674–13678. https://doi.org/10.1073/pnas.96.24.13674
346. Honig, B.; Yang, A.-S. Free Energy Balance in Protein Folding. In *Adv. Protein Chem.*; Anfinsen, C. B.; Richards, F. M.; Edsall, J. T.; Eisenberg,

D. S., Eds.; Academic Press: Protein Stability, Vol. 46, 1995; pp 27–58. https://doi.org/10.1016/S0065-3233(08)60331-9

347. Xiao, L.; Honig, B. Electrostatic Contributions to the Stability of Hyperthermophilic Proteins. *J. Mol. Biol.* **1999**, *289* (5), 1435–1444. https://doi.org/10.1006/jmbi.1999.2810
348. Giletto, A.; Pace, C. N. Buried, Charged, Non-Ion-Paired Aspartic Acid 76 Contributes Favorably to the Conformational Stability of Ribonuclease T1. *Biochemistry* **1999**, *38* (40), 13379–13384. https://doi.org/10.1021/bi991422s
349. Tanford, C.; Kirkwood, J. G. Theory of Protein Titration Curves. I. General Equations for Impenetrable Spheres. *J. Am. Chem. Soc.* **1957**, *79* (20), 5333–5339. https://doi.org/10.1021/ja01577a001
350. Warshel, A.; Russell, S. T.; Churg, A. K. Macroscopic Models for Studies of Electrostatic Interactions in Proteins: Limitations and Applicability. *Proc. Natl. Acad. Sci. U.S.A.* **1984**, *81* (15), 4785–4789. https://doi.org/10.1073/pnas.81.15.4785
351. Roca, M.; Messer, B.; Warshel, A. Electrostatic Contributions to Protein Stability and Folding Energy. *FEBS Lett.* **2007**, *581* (10), 2065–2071. https://doi.org/10.1016/j.febslet.2007.04.025
352. Vicatos, S.; Roca, M.; Warshel, A. Effective Approach for Calculations of Absolute Stability of Proteins Using Focused Dielectric Constants. *Proteins* **2009**, *77* (3), 670–684. https://doi.org/10.1002/prot.22481
353. Hille, B. *Ion Channels of Excitable Membranes*; Sinauer Assoc.: Sunderland, MA, 2001.
354. Eisenman, G.; Horn, R. Ionic Selectivity Revisited: The Role of Kinetic and Equilibrium Processes in Ion Permeation through Channels. *J. Membr. Biol.* **1983**, *76* (3), 197–225. https://doi.org/10.1007/BF01870364
355. Jordan, P. C. Microscopic Approaches to Ion Transport through Transmembrane Channels: The Model System Gramicidin. *J. Phys. Chem.* **1987**, *91* (27), 6582–6591. https://doi.org/10.1021/j100311a006

356. Aqvist, J.; Warshel, A. Energetics of Ion Permeation through Membrane Channels. Solvation of Na^+ by Gramicidin A. *Biophys. J.* **1989**, *56* (1), 171–182. https://doi.org/10.1016/S0006-3495(89)82662-1
357. Morais-Cabral, J. H.; Zhou, Y.; MacKinnon, R. Energetic Optimization of Ion Conduction Rate by the K+ Selectivity Filter. *Nature* **2001**, *414* (6859), 37–42. https://doi.org/10.1038/35102000
358. Doyle, D. A.; Cabral, J. M.; Pfuetzner, R. A.; Kuo, A.; Gulbis, J. M.; Cohen, S. L.; Chait, B. T.; MacKinnon, R. The Structure of the Potassium Channel: Molecular Basis of K^+ Conduction and Selectivity. *Science* **1998**, *280* (5360), 69–77. https://doi.org/10.1126/science.280.5360.69
359. Shrivastava, I. H.; Sansom, M. S. P. Simulations of Ion Permeation Through a Potassium Channel: Molecular Dynamics of KcsA in a Phospholipid Bilayer. *Biophys. J.* **2000**, *78* (2), 557–570. https://doi.org/10.1016/S0006-3495(00)76616-1
360. Allen, T. W.; Kuyucak, S.; Chung, S.-H. Molecular Dynamics Study of the KcsA Potassium Channel. *Biophys. J.* **1999**, *77* (5), 2502–2516. https://doi.org/10.1016/S0006-3495(99)77086-4
361. Åqvist, J.; Luzhkov, V. Ion Permeation Mechanism of the Potassium Channel. *Nature* **2000**, *404* (6780), 881–884. https://doi.org/10.1038/35009114
362. Dryga, A.; Chakrabarty, S.; Vicatos, S.; Warshel, A. Realistic Simulation of the Activation of Voltage-Gated Ion Channels. *Proc. Natl. Acad. Sci. U.S.A.* **2012**, *109* (9), 3335–3340. https://doi.org/10.1073/pnas.1121094100
363. Dryga, A.; Chakrabarty, S.; Vicatos, S.; Warshel, A. Coarse Grained Model for Exploring Voltage Dependent Ion Channels. *BBA – Biomembranes* **2012**, *1818* (2), 303–317. https://doi.org/10.1016/j.bbamem.2011.07.043
364. Lee, M.; Kolev, V.; Warshel, A. Validating a Coarse-Grained Voltage Activation Model by Comparing Its Performance to the Results of Monte Carlo Simulations. *J. Phys. Chem. B* **2017**, *121* (50), 11284–11291. https://doi.org/10.1021/acs.jpcb.7b09530

365. Delemotte, L.; Tarek, M.; Klein, M. L.; Amaral, C.; Treptow, W. Intermediate States of the Kv1.2 Voltage Sensor from Atomistic Molecular Dynamics Simulations. *Proc. Natl. Acad. Sci. U.S.A.* **2011**, *108* (15), 6109–6114. https://doi.org/10.1073/pnas.1102724100
366. Vargas, E.; Bezanilla, F.; Roux, B. In Search of a Consensus Model of the Resting State of a Voltage-Sensing Domain. *Neuron* **2011**, 72 (5), 713–720. https://doi.org/10.1016/j.neuron.2011.09.024
367. Khalili-Araghi, F.; Tajkhorshid, E.; Roux, B.; Schulten, K. Molecular Dynamics Investigation of the ω-Current in the Kv1.2 Voltage Sensor Domains. *Biophys. J.* **2012**, *102* (2), 258–267. https://doi.org/10.1016/j.bpj.2011.10.057
368. Jensen, M. Ø.; Jogini, V.; Borhani, D. W.; Leffler, A. E.; Dror, R. O.; Shaw, D. E. Mechanism of Voltage Gating in Potassium Channels. *Science* **2012**, *336* (6078), 229–233. https://doi.org/10.1126/science.1216538
369. Treptow, W.; Klein, M. L. Computer Simulations of Voltage-Gated Cation Channels. *J. Phys. Chem. Lett.* **2012**, *3* (8), 1017–1023. https://doi.org/10.1021/jz300089g
370. Freites, J. A.; Schow, E. V.; White, S. H.; Tobias, D. J. Microscopic Origin of Gating Current Fluctuations in a Potassium Channel Voltage Sensor. *Biophys. J.* **2012**, *102* (11), L44–L46. https://doi.org/10.1016/j.bpj.2012.04.021
371. Roux, B. Statistical Mechanical Equilibrium Theory of Selective Ion Channels. *Biophys. J.* **1999**, 77 (1), 139–153. https://doi.org/10.1016/S0006-3495(99)76878-5
372. Roux, B. The Membrane Potential and Its Representation by a Constant Electric Field in Computer Simulations. *Biophys. J.* **2008**, *95* (9), 4205–4216. https://doi.org/10.1529/biophysj.108.136499
373. Grabe, M.; Lai, H. C.; Jain, M.; Nung Jan, Y.; Yeh Jan, L. Structure Prediction for the Down State of a Potassium Channel Voltage Sensor. *Nature* **2007**, *445* (7127), 550–553. https://doi.org/10.1038/nature05494
374. Grabe, M.; Lecar, H.; Jan, Y. N.; Jan, L. Y. A Quantitative Assessment of Models for Voltage-Dependent Gating of Ion Channels. *Proc. Natl. Acad.*

Sci. U.S.A. **2004**, *101* (51), 17640–17645. https://doi.org/10.1073/pnas.0408116101

375. Parsegian, A. Energy of an Ion Crossing a Low Dielectric Membrane: Solutions to Four Relevant Electrostatic Problems. *Nature* **1969**, *221* (5183), 844–846. https://doi.org/10.1038/221844a0
376. Läuger, P. Ion Transport through Pores: A Rate-Theory Analysis. *BBA — Biomembranes* **1973**, *311* (3), 423–441. https://doi.org/10.1016/0005-2736(73)90323-4
377. Torrie, G. M.; Valleau, J. P. Electrical Double Layers. I. Monte Carlo Study of a Uniformly Charged Surface. *J. Chem. Phys.* **1980**, *73* (11), 5807–5816. https://doi.org/10.1063/1.440065
378. Kim, I.; Chakrabarty, S.; Brzezinski, P.; Warshel, A. Modeling Gating Charge and Voltage Changes in Response to Charge Separation in Membrane Proteins. *Proc. Natl. Acad. Sci. U.S.A.* **2014**, *111* (31), 11353–11358. https://doi.org/10.1073/pnas.1411573111
379. Kim, I.; Warshel, A. Analyzing the Electrogenicity of Cytochrome c Oxidase. *Proc. Natl. Acad. Sci. U.S.A.* **2016**, *113* (28), 7810–7815. https://doi.org/10.1073/pnas.1608118113
380. Ostermeier, C.; Harrenga, A.; Ermler, U.; Michel, H. Structure at 2.7 Å Resolution of the Paracoccus Denitrificans Two-Subunit Cytochrome c Oxidase Complexed with an Antibody FV Fragment. *Proc. Natl. Acad. Sci. U.S.A.* **1997**, *94* (20), 10547–10553. https://doi.org/10.1073/pnas.94.20.10547
381. Yoshikawa, S.; Shinzawa-Itoh, K.; Nakashima, R.; Yaono, R.; Yamashita, E.; Inoue, N.; Yao, M.; Fei, M. J.; Libeu, C. P.; Mizushima, T.; Yamaguchi, H.; Tomizaki, T.; Tsukihara, T. Redox-Coupled Crystal Structural Changes in Bovine Heart Cytochrome c Oxidase. *Science* **1998**, *280* (5370), 1723–1729. https://doi.org/10.1126/science.280.5370.1723
382. Luecke, H.; Schobert, B.; Richter, H.-T.; Cartailler, J.-P.; Lanyi, J. K. Structure of Bacteriorhodopsin at 1.55 Å Resolution. *J. Mol. Biol.* **1999**, *291* (4), 899–911. https://doi.org/10.1006/jmbi.1999.3027

383. Sass, H. J.; Büldt, G.; Gessenich, R.; Hehn, D.; Neff, D.; Schlesinger, R.; Berendzen, J.; Ormos, P. Structural Alterations for Proton Translocation in the M State of Wild-Type Bacteriorhodopsin. *Nature* **2000**, *406* (6796), 649–653. https://doi.org/10.1038/35020607
384. Royant, A.; Edman, K.; Ursby, T.; Pebay-Peyroula, E.; Landau, E. M.; Neutze, R. Helix Deformation Is Coupled to Vectorial Proton Transport in the Photocycle of Bacteriorhodopsin. *Nature* **2000**, *406* (6796), 645–648. https://doi.org/10.1038/35020599
385. Luecke, H. Atomic Resolution Structures of Bacteriorhodopsin Photocycle Intermediates: The Role of Discrete Water Molecules in the Function of This Light-Driven Ion Pump. *BBA – Bioenergetics* **2000**, *1460* (1), 133–156. https://doi.org/10.1016/S0005-2728(00)00135-3
386. Lanyi, J. K. Bacteriorhodopsin. *Annu. Rev. Physiol.* **2004**, *66*, 665–688. https://doi.org/10.1146/annurev.physiol.66.032102.150049
387. Meier, T.; Polzer, P.; Diederichs, K.; Welte, W.; Dimroth, P. Structure of the Rotor Ring of F-Type Na^+-ATPase from Ilyobacter tartaricus. *Science* **2005**, *308* (5722), 659–662. https://doi.org/10.1126/science.1111199
388. Murata, T.; Yamato, I.; Kakinuma, Y.; Leslie, A. G. W.; Walker, J. E. Structure of the Rotor of the V-Type Na^+-ATPase from Enterococcus hirae. *Science* **2005**, *308* (5722), 654–659. https://doi.org/10.1126/science.1110064
389. Nishida, M.; Cadene, M.; Chait, B. T.; MacKinnon, R. Crystal Structure of a Kir3.1–Prokaryotic Kir Channel Chimera. *The EMBO J.* **2007**, *26* (17), 4005–4015. https://doi.org/10.1038/sj.emboj.7601828
390. Cordero-Morales, J. F.; Jogini, V.; Lewis, A.; Vásquez, V.; Cortes, D. M.; Roux, B.; Perozo, E. Molecular Driving Forces Determining Potassium Channel Slow Inactivation. *Nat. Struct. Mol. Biol.* **2007**, *14* (11), 1062–1069. https://doi.org/10.1038/nsmb1309
391. Long, S. B.; Campbell, E. B.; MacKinnon, R. Crystal Structure of a Mammalian Voltage-Dependent Shaker Family K^+ Channel. *Science* **2005**, *309* (5736), 897–903. https://doi.org/10.1126/science.1116269

392. Bichet, D.; Grabe, M.; Jan, Y. N.; Jan, L. Y. Electrostatic Interactions in the Channel Cavity as an Important Determinant of Potassium Channel Selectivity. *Proc. Natl. Acad. Sci. U.S.A.* **2006**, *103* (39), 14355–14360. https://doi.org/10.1073/pnas.0606660103
393. Agre, P.; Kozono, D. Aquaporin Water Channels: Molecular Mechanisms for Human Diseases. Presented at the Nobel Symposium 126, Membrane Proteins: Structure, Function, and Assembly, Friibergh's Herrgård, Örsundsbro, Sweden, August 23, 2003. *FEBS Lett.* **2003**, *555* (1), 72–78. https://doi.org/10.1016/S0014-5793(03)01083-4
394. Silverman, D. N.; Tu, C.; Chen, X.; Tanhauser, S. M.; Kresge, A. J.; Laipis, P. J. Rate-Equilibria Relationships in Intramolecular Proton Transfer in Human Carbonic Anhydrase III. *Biochemistry* **1993**, *32* (40), 10757–10762. https://doi.org/10.1021/bi00091a029
395. Schutz, C. N.; Warshel, A. Analyzing Free Energy Relationships for Proton Translocations in Enzymes: Carbonic Anhydrase Revisited. *J. Phys. Chem. B* **2004**, *108* (6), 2066–2075. https://doi.org/10.1021/jp0366015
396. Braun-Sand, S.; Strajbl, M.; Warshel, A. Studies of Proton Translocations in Biological Systems: Simulating Proton Transport in Carbonic Anhydrase by EVB-Based Models. *Biophys. J.* **2004**, *87* (4), 2221–2239. https://doi.org/10.1529/biophysj.104.043257
397. Riccardi, D.; König, P.; Guo, H.; Cui, Q. Proton Transfer in Carbonic Anhydrase Is Controlled by Electrostatics Rather Than the Orientation of the Acceptor. *Biochemistry* **2008**, *47* (8), 2369–2378. https://doi.org/10.1021/bi701950j
398. Riccardi, D.; König, P.; Prat-Resina, X.; Yu, H.; Elstner, M.; Frauenheim, T.; Cui, Q. "Proton Holes" in Long-Range Proton Transfer Reactions in Solution and Enzymes: A Theoretical Analysis. *J. Am. Chem. Soc.* **2006**, *128* (50), 16302–16311. https://doi.org/10.1021/ja065451j
399. Braun-Sand, S.; Burykin, A.; Chu, Z. T.; Warshel, A. Realistic Simulations of Proton Transport Along the Gramicidin Channel: Demonstrating the

Importance of Solvation Effects. *J. Phys. Chem. B* **2005**, *109* (1), 583–592. https://doi.org/10.1021/jp0465783

400. Sham, Y. Y.; Muegge, I.; Warshel, A. Simulating Proton Translocations in Proteins: Probing Proton Transfer Pathways in the Rhodobacter sphaeroides Reaction Center. *Proteins* **1999**, *36* (4), 484–500. https://doi.org/10.1002/(SICI)1097-0134(19990901)36:4<484::AID-PROT13>3.0.CO;2-R
401. Pisliakov, A. V.; Sharma, P. K.; Chu, Z. T.; Haranczyk, M.; Warshel, A. Electrostatic Basis for the Unidirectionality of the Primary Proton Transfer in Cytochrome c Oxidase. *Proc. Natl. Acad. Sci. U.S.A.* **2008**, *105* (22), 7726–7731. https://doi.org/10.1073/pnas.0800580105
402. Murata, K.; Mitsuoka, K.; Hirai, T.; Walz, T.; Agre, P.; Heymann, J. B.; Engel, A.; Fujiyoshi, Y. Structural Determinants of Water Permeation through Aquaporin-1. *Nature* **2000**, *407* (6804), 599–605. https://doi.org/10.1038/35036519
403. Sui, H.; Han, B.-G.; Lee, J. K.; Walian, P.; Jap, B. K. Structural Basis of Water-Specific Transport through the AQP1 Water Channel. *Nature* **2001**, *414* (6866), 872–878. https://doi.org/10.1038/414872a
404. Yarnell, A. Blockade in the Cell's Waterway. *Chem. Eng. News.* accessed December 23, 2023.
405. de Groot, B. L.; Grubmüller, H. Water Permeation Across Biological Membranes: Mechanism and Dynamics of Aquaporin-1 and GlpF. *Science* **2001**, *294* (5550), 2353–2357. https://doi.org/10.1126/science.1066115
406. de Groot, B. L.; Frigato, T.; Helms, V.; Grubmüller, H. The Mechanism of Proton Exclusion in the Aquaporin-1 Water Channel. *J. Mol. Biol.* **2003**, *333* (2), 279–293. https://doi.org/10.1016/j.jmb.2003.08.003
407. Decoursey, T. E. Voltage-Gated Proton Channels and Other Proton Transfer Pathways. *Physiol. Rev.* **2003**, *83* (2), 475–579. https://doi.org/10.1152/physrev.00028.2002
408. Jensen, M. Ø.; Tajkhorshid, E.; Schulten, K. Electrostatic Tuning of Permeation and Selectivity in Aquaporin Water Channels. *Biophys. J.* **2003**, *85* (5), 2884–2899. https://doi.org/10.1016/S0006-3495(03)74711-0

409. Tajkhorshid, E.; Nollert, P.; Jensen, M. Ø.; Miercke, L. J. W.; O'Connell, J.; Stroud, R. M.; Schulten, K. Control of the Selectivity of the Aquaporin Water Channel Family by Global Orientational Tuning. *Science* **2002**, *296* (5567), 525–530. https://doi.org/10.1126/science.1067778
410. Burykin, A.; Warshel, A. On the Origin of the Electrostatic Barrier for Proton Transport in Aquaporin. *FEBS Lett.* **2004**, *570* (1), 41–46. https://doi.org/10.1016/j.febslet.2004.06.020
411. Burykin, A.; Warshel, A. What Really Prevents Proton Transport through Aquaporin? Charge Self-Energy versus Proton Wire Proposals. *Biophys. J.* **2003**, *85* (6), 3696–3706. https://doi.org/10.1016/S0006-3495(03)74786-9
412. Miloshevsky, G. V.; Jordan, P. C. Water and Ion Permeation in bAQP1 and GlpF Channels: A Kinetic Monte Carlo Study. *Biophys. J.* **2004**, *87* (6), 3690–3702. https://doi.org/10.1529/biophysj.104.043315
413. Pasqualato, S.; Cherfils, J. Crystallographic Evidence for Substrate-Assisted GTP Hydrolysis by a Small GTP Binding Protein. *Structure* **2005**, *13* (4), 533–540. https://doi.org/10.1016/j.str.2005.01.014
414. Sprang, S. R. G PROTEIN MECHANISMS: Insights from Structural Analysis. *Annu. Rev. Biochem.* **1997**, *66* (1), 639–678. https://doi.org/10.1146/annurev.biochem.66.1.639
415. de Groot, B. L.; Grubmüller, H. The Dynamics and Energetics of Water Permeation and Proton Exclusion in Aquaporins. *Curr. Opin. Struct. Biol.* **2005**, *15* (2), 176–183. https://doi.org/10.1016/j.sbi.2005.02.003
416. Florián, J.; Warshel, A. A Fundamental Assumption about OH– Attack in Phosphate Ester Hydrolysis Is Not Fully Justified. *J. Am. Chem. Soc.* **1997**, *119* (23), 5473–5474. https://doi.org/10.1021/ja964270m
417. Nagle, J. F.; Mille, M.; Morowitz, H. J. Theory of Hydrogen Bonded Chains in Bioenergetics. *J. Chem. Phys.* **1980**, 72 (7), 3959–3971. https://doi.org/10.1063/1.439674
418. Nagle, J. F.; Mille, M. Molecular Models of Proton Pumps. *J. Chem. Phys.* **1981**, *74* (2), 1367–1372. https://doi.org/10.1063/1.441199

419. Ramsey, I. S.; Mokrab, Y.; Carvacho, I.; Sands, Z. A.; Sansom, M. S. P.; Clapham, D. E. An Aqueous H^+ Permeation Pathway in the Voltage-Gated Proton Channel Hv1. *Nat. Struct. Mol. Biol.* **2010**, *17* (7), 869–875. https://doi.org/10.1038/nsmb.1826
420. Smith, S. M. E.; Morgan, D.; Musset, B.; Cherny, V. V.; Place, A. R.; Hastings, J. W.; DeCoursey, T. E. Voltage-Gated Proton Channel in a Dinoflagellate. *Proc. Natl. Acad. Sci. U.S.A.* **2011**, *108* (44), 18162–18167. https://doi.org/10.1073/pnas.1115405108
421. Berger, T. K.; Isacoff, E. Y. The Pore of the Voltage-Gated Proton Channel. *Neuron* **2011**, *72* (6), 991–1000. https://doi.org/10.1016/j.neuron.2011.11.014
422. Musset, B.; Smith, S. M. E.; Rajan, S.; Morgan, D.; Cherny, V. V.; DeCoursey, T. E. Aspartate 112 Is the Selectivity Filter of the Human Voltage-Gated Proton Channel. *Nature* **2011**, *480* (7376), 273–277. https://doi.org/10.1038/nature10557
423. Pupo, A.; Baez-Nieto, D.; Martínez, A.; Latorre, R.; González, C. Proton Channel Models: Filling the Gap between Experimental Data and the Structural Rationale. *Channels* **2014**, *8* (3), 180–192. https://doi.org/10.4161/chan.28665
424. DeCoursey, T. E. The Voltage-Gated Proton Channel: A Riddle, Wrapped in a Mystery, inside an Enigma. *Biochemistry* **2015**, *54* (21), 3250–3268. https://doi.org/10.1021/acs.biochem.5b00353
425. Takeshita, K.; Sakata, S.; Yamashita, E.; Fujiwara, Y.; Kawanabe, A.; Kurokawa, T.; Okochi, Y.; Matsuda, M.; Narita, H.; Okamura, Y.; Nakagawa, A. X-Ray Crystal Structure of Voltage-Gated Proton Channel. *Nat. Struct. Mol. Biol.* **2014**, *21* (4), 352–357. https://doi.org/10.1038/nsmb.2783
426. Okamura, Y.; Fujiwara, Y.; Sakata, S. Gating Mechanisms of Voltage-Gated Proton Channels. *Annu. Rev. Biochem.* **2015**, *84*, 685–709. https://doi.org/10.1146/annurev-biochem-060614-034307

427. Freites, J. A.; Tobias, D. J. Voltage Sensing in Membranes: From Macroscopic Currents to Molecular Motions. *J. Membr. Biol.* **2015**, *248* (3), 419–430. https://doi.org/10.1007/s00232-015-9805-x

428. Geragotelis, A. D.; Wood, M. L.; Göddeke, H.; Hong, L.; Webster, P. D.; Wong, E. K.; Freites, J. A.; Tombola, F.; Tobias, D. J. Voltage-Dependent Structural Models of the Human Hv1 Proton Channel from Long-Timescale Molecular Dynamics Simulations. *Proc. Natl. Acad. Sci. U.S.A.* **2020**, *117* (24), 13490–13498. https://doi.org/10.1073/pnas.1920943117

429. Li, Q.; Shen, R.; Treger, J. S.; Wanderling, S. S.; Milewski, W.; Siwowska, K.; Bezanilla, F.; Perozo, E. Resting State of the Human Proton Channel Dimer in a Lipid Bilayer. *Proc. Natl. Acad. Sci. U.S.A.* **2015**, *112* (44), E5926–E5935. https://doi.org/10.1073/pnas.1515043112

430. Dudev, T.; Musset, B.; Morgan, D.; Cherny, V. V.; Smith, S. M. E.; Mazmanian, K.; DeCoursey, T. E.; Lim, C. Selectivity Mechanism of the Voltage-Gated Proton Channel, HV1. *Sci. Rep.* **2015**, *5*, 10320. https://doi.org/10.1038/srep10320

431. Chamberlin, A.; Qiu, F.; Wang, Y.; Noskov, S. Y.; Peter Larsson, H. Mapping the Gating and Permeation Pathways in the Voltage-Gated Proton Channel Hv1. *J. Mol. Biol.* **2015**, *427* (1), 131–145. https://doi.org/10.1016/j.jmb.2014.11.018

432. Lee, M.; Bai, C.; Feliks, M.; Alhadeff, R.; Warshel, A. On the Control of the Proton Current in the Voltage-Gated Proton Channel Hv1. *Proc. Natl. Acad. Sci. U.S.A.* **2018**, *115* (41), 10321–10326. https://doi.org/10.1073/pnas.1809766115

433. Kim, I.; Warshel, A. A Microscopic Capacitor Model of Voltage Coupling in Membrane Proteins: Gating Charge Fluctuations in Ci-VSD. *J. Phys. Chem. B* **2016**, *120* (3), 418–432. https://doi.org/10.1021/acs.jpcb.5b10956

434. Dryga, A.; Chakrabarty, S.; Vicatos, S.; Warshel, A. Coarse Grained Model for Exploring Voltage Dependent Ion Channels. *BBA – Biomembranes* **2012**, *1818* (2), 303–317. https://doi.org/10.1016/j.bbamem.2011.07.043

435. Kim, I.; Warshel, A. Coarse-Grained Simulations of the Gating Current in the Voltage-Activated Kv1.2 Channel. *Proc. Natl. Acad. Sci. U.S.A.* **2014**, *111* (6), 2128–2133. https://doi.org/10.1073/pnas.1324014111

436. Wood, M. L.; Schow, E. V.; Freites, J. A.; White, S. H.; Tombola, F.; Tobias, D. J. Water Wires in Atomistic Models of the Hv1 Proton Channel. *BBA – Biomembranes* **2012**, *1818* (2), 286–293. https://doi.org/10.1016/j.bbamem.2011.07.045

437. van Keulen, S. C.; Gianti, E.; Carnevale, V.; Klein, M. L.; Rothlisberger, U.; Delemotte, L. Does Proton Conduction in the Voltage-Gated H^+ Channel hHv1 Involve Grotthuss-Like Hopping via Acidic Residues? *J. Phys. Chem. B* **2017**, *121* (15), 3340–3351. https://doi.org/10.1021/acs.jpcb.6b08339

438. Nassar, N.; Horn, G.; Herrmann, C. A.; Scherer, A.; McCormick, F.; Wittinghofer, A. The 2.2 Å Crystal Structure of the Ras-Binding Domain of the Serine/Threonine Kinase c-Raf1 in Complex with RaplA and a GTP Analogue. *Nature* **1995**, *375* (6532), 554–560. https://doi.org/10.1038/375554a0

439. Menz, R. I.; Walker, J. E.; Leslie, A. G. W. Structure of Bovine Mitochondrial F_1-ATPase with Nucleotide Bound to All Three Catalytic Sites: Implications for the Mechanism of Rotary Catalysis. *Cell* **2001**, *106* (3), 331–341. https://doi.org/10.1016/S0092-8674(01)00452-4

440. Victor, B. L.; Vicente, J. B.; Rodrigues, R.; Oliveira, S.; Rodrigues-Pousada, C.; Frazão, C.; Gomes, C. M.; Teixeira, M.; Soares, C. M. Docking and Electron Transfer Studies between Rubredoxin and Rubredoxin:Oxygen Oxidoreductase. *J. Biol. Inorg. Chem.* **2003**, *8* (4), 475–488. https://doi.org/10.1007/s00775-002-0440-5

441. Flöck, D.; Helms, V. Protein–Protein Docking of Electron Transfer Complexes: Cytochrome c Oxidase and Cytochrome c. *Proteins* **2002**, *47* (1), 75–85. https://doi.org/10.1002/prot.10066

442. Bai, C.; Wang, J.; Chen, G.; Zhang, H.; An, K.; Xu, P.; Du, Y.; Ye, R. D.; Saha, A.; Zhang, A.; Warshel, A. Predicting Mutational Effects on Receptor

Binding of the Spike Protein of SARS-CoV-2 Variants. *J. Am. Chem. Soc.* **2021**, *143* (42), 17646–17654. https://doi.org/10.1021/jacs.1c07965

443. Coleman, D. E.; Sprang, S. R. [4] Reaction Dynamics of G-Protein Catalyzed Hydrolysis of GTP as Viewed by X-Ray Crystallographic Snapshots of Giα1. In *Methods in Enzymology, Enzyme Kinetics and Mechanism Part E: Energetics of Enzyme Catalysis*; Academic Press, 1999; Vol. 308, pp 70–92. https://doi.org/10.1016/S0076-6879(99)08006-4
444. Scheffzek, K.; Ahmadian, M. R. GTPase Activating Proteins: Structural and Functional Insights 18 Years after Discovery. *Cell. Mol. Life Sci.* **2005**, *62* (24), 3014–3038. https://doi.org/10.1007/s00018-005-5136-x
445. Sprang, S. R. G Proteins, Effectors and GAPs: Structure and Mechanism. *Curr. Opin. Struct. Biol.* **1997**, *7* (6), 849–856. https://doi.org/10.1016/S0959-440X(97)80157-1
446. Vetter, I. R.; Wittinghofer, A. Nucleoside Triphosphate-Binding Proteins: Different Scaffolds to Achieve Phosphoryl Transfer. *Quat. Rev. Biophys.* **1999**, *32* (1), 1–56. https://doi.org/10.1017/S0033583599003480
447. Wittinghofer, A. Phosphoryl Transfer in Ras Proteins, Conclusive or Elusive? *Trends Biochem. Sci.* **2006**, *31* (1), 20–23. https://doi.org/10.1016/j.tibs.2005.11.012
448. Barbacid, M. Ras GENES. *Annu. Rev. Biochem.* **1987**, *56* (1), 779–827. https://doi.org/10.1146/annurev.bi.56.070187.004023
449. Bos, J. L. Ras Oncogenes in Human Cancer: A Review. *Cancer Res.* **1989**, *49* (17), 4682–4689.
450. Bourne, H. R.; Sanders, D. A.; McCormick, F. The GTPase Superfamily: Conserved Structure and Molecular Mechanism. *Nature* **1991**, *349* (6305), 117–127. https://doi.org/10.1038/349117a0
451. Scheffzek, K.; Ahmadian, M. R.; Kabsch, W.; Wiesmüller, L.; Lautwein, A.; Schmitz, F.; Wittinghofer, A. The Ras–RasGAP Complex: Structural Basis for GTPase Activation and Its Loss in Oncogenic Ras Mutants. *Science* **1997**, *277* (5324), 333–339. https://doi.org/10.1126/science.277.5324.333

452. Pai, E. F.; Krengel, U.; Petsko, G. A.; Goody, R. S.; Kabsch, W.; Wittinghofer, A. Refined Crystal Structure of the Triphosphate Conformation of H-ras P21 at 1.35 Å Resolution: Implications for the Mechanism of GTP Hydrolysis. *EMBO J.* **1990**, *9* (8), 2351–2359. https://doi.org/10.1002/j.1460-2075.1990.tb07409.x

453. Krengel, U.; Schlichting, I.; Scherer, A.; Schumann, R.; Frech, M.; John, J.; Kabsch, W.; Pai, E. F.; Wittinghofer, A. Three-Dimensional Structures of H-Ras P21 Mutants: Molecular Basis for Their Inability to Function as Signal Switch Molecules. *Cell* **1990**, *62* (3), 539–548. https://doi.org/10.1016/0092-8674(90)90018-A

454. Langen, R.; Schweins, T.; Warshel, A. On the Mechanism of Guanosine Triphosphate Hydrolysis in Ras p21 Proteins. *Biochemistry* **1992**, *31*, 8691–8696. https://doi.org/10.1021/bi00152a002

455. Schweins, T.; Geyer, M.; Scheffzek, K.; Warshel, A.; Kalbitzer, H. R.; Wittinghofer, A. Substrate-Assisted Catalysis as a Mechanism for GTP Hydrolysis of P21ras and Other GTP-Binding Proteins. *Nat. Struct. Mol. Biol.* **1995**, 2 (1), 36–44. https://doi.org/10.1038/nsb0195-36

456. Schweins, T.; Langen, R.; Warshel, A. Why Have Mutagenesis Studies Not Located the General Base in Ras p21. *Nat. Struct. Mol. Biol.* **1994**, *1* (7), 476–484. https://doi.org/10.1038/nsb0794-476

457. Kamerlin, S. C. L.; Sharma, P. K.; Prasad, R. B.; Warshel, A. Why Nature Really Chose Phosphate. *Quat. Rev. Biophys.* **2013**, *46* (1), 1–132. https://doi.org/10.1017/S0033583512000157

458. Chung, H.-H.; Benson, D. R.; Schultz, P. G. Probing the Structure and Mechanism of Ras Protein with an Expanded Genetic Code. *Science* **1993**, *259* (5096), 806–809. https://doi.org/10.1126/science.8430333

459. Dall'acqua, W.; Carter, P. Substrate-Assisted Catalysis: Molecular Basis and Biological Significance. *Protein Sci.* **2000**, *9* (1), 1–9. https://doi.org/10.1110/ps.9.1.1

460. Kosloff, M.; Zor, T.; Selinger, Z. Substrate-Assisted Catalysis: Implications for Biotechnology and Drug Design. *Drug Dev. Res.* **2000**, *50* (3–4), 250–257. https://doi.org/10.1002/1098-2299(200007/08)50:3/4<250::AID-DDR8>3.0.CO;2-0
461. Klähn, M.; Rosta, E.; Warshel, A. On the Mechanism of Hydrolysis of Phosphate Monoesters Dianions in Solutions and Proteins. *J. Am. Chem. Soc.* **2006**, *128* (47), 15310–15323. https://doi.org/10.1021/ja065470t
462. Kamerlin, S. C. L.; Williams, N. H.; Warshel, A. Dineopentyl Phosphate Hydrolysis: Evidence for Stepwise Water Attack. *J. Org. Chem.* **2008**, *73* (18), 6960–6969. https://doi.org/10.1021/jo801207q
463. Schweins, T.; Geyer, M.; Kalbitzer, H. R.; Wittinghofer, A.; Warshel, A. Linear Free Energy Relationships in the Intrinsic and GTPase Activating Protein-Stimulated Guanosine 5′-Triphosphate Hydrolysis of P21ras. *Biochemistry* **1996**, *35* (45), 14225–14231. https://doi.org/10.1021/bi961118o
464. Grigorenko, B. L.; Nemukhin, A. V.; Topol, I. A.; Cachau, R. E.; Burt, S. K. QM/MM Modeling the Ras–GAP Catalyzed Hydrolysis of Guanosine Triphosphate. *Proteins* **2005**, *60* (3), 495–503. https://doi.org/10.1002/prot.20472
465. Grigorenko, B. L.; Rogov, A. V.; Nemukhin, A. V. Mechanism of Triphosphate Hydrolysis in Aqueous Solution: QM/MM Simulations in Water Clusters. *J. Phys. Chem. B* **2006**, *110* (9), 4407–4412. https://doi.org/10.1021/jp056395w
466. Plotnikov, N. V.; Prasad, B. R.; Lameira, J.; Burt, S. K.; Warshel, A. Quantitative Exploration of the Molecular Origin of the Activation of GTPase. *Proc. Natl. Acad. Sci. U.S.A.* **2013**, *110* (51), 20509–20514. https://doi.org/10.1073/pnas.1319854110
467. Prasad, B. R.; Plotnikov, N. V.; Warshel, A. Addressing Open Questions about Phosphate Hydrolysis Pathways by Careful Free Energy Mapping. *J. Phys. Chem. B* **2013**, *117* (1), 153–163. https://doi.org/10.1021/jp309778n

468. Plotnikov, N. V.; Prasad, B. R.; Chakrabarty, S.; Chu, Z. T.; Warshel, A. Quantifying the Mechanism of Phosphate Monoester Hydrolysis in Aqueous Solution by Evaluating the Relevant Ab Initio QM/MM Free-Energy Surfaces. *J. Phys. Chem. B* **2013**, *117* (42), 12807–12819. https://doi.org/10.1021/jp4020146
469. Daviter, T.; Wieden, H.-J.; Rodnina, M. V. Essential Role of Histidine 84 in Elongation Factor Tu for the Chemical Step of GTP Hydrolysis on the Ribosome. *J. Mol. Biol.* **2003**, *332* (3), 689–699. https://doi.org/10.1016/S0022-2836(03)00947-1
470. Schweins, T. Untersuchung des Reaktionsmechanismus von Ras-P21 mittels Computer-Modeling; Diplomarbeit thesis, University of Southern California, 1991.
471. Graziano, M. P.; Gilman, A. G. Synthesis in Escherichia Coli of GTPase-Deficient Mutants of Gsα*. *J. Biol. Chem.* **1989**, *264* (26), 15475–15482. https://doi.org/10.1016/S0021-9258(89)84854-8
472. Allin, C.; Gerwert, K. Ras Catalyzes GTP Hydrolysis by Shifting Negative Charges from γ- to β-Phosphate As Revealed by Time-Resolved FTIR Difference Spectroscopy. *Biochemistry* **2001**, *40* (10), 3037–3046. https://doi.org/10.1021/bi0017024
473. Allin, C.; Ahmadian, M. R.; Wittinghofer, A.; Gerwert, K. Monitoring the GAP Catalyzed H-Ras GTPase Reaction at Atomic Resolution in Real Time. *Proc. Natl. Acad. Sci. U.S.A.* **2001**, *98* (14), 7754–7759. https://doi.org/10.1073/pnas.131549798
474. Kawashima, T.; Berthet-Colominas, C.; Wulff, M.; Cusack, S.; Leberman, R. The Structure of the Escherichia Coli EF-Tu·EF-Ts Complex at 2.5 Å Resolution. *Nature* **1996**, *379* (6565), 511–518. https://doi.org/10.1038/379511a0
475. Wang, Y.; Jiang, Y.; Meyering-Voss, M.; Sprinzl, M.; Sigler, P. B. Crystal Structure of the EF-Tu·EF-Ts Complex from *Thermus thermophilus*. *Nat. Struct. Mol. Biol.* **1997**, *4* (8), 650–656. https://doi.org/10.1038/nsb0897-650

476. Maracci, C.; Peske, F.; Dannies, E.; Pohl, C.; Rodnina, M. V. Ribosome-Induced Tuning of GTP Hydrolysis by a Translational GTPase. *Proc. Natl. Acad. Sci. U.S.A.* **2014**, *111* (40), 14418–14423. https://doi.org/10.1073/pnas.1412676111

477. Liljas, A. *Structural Aspects of Protein Synthesis*; World Scientific Publishing Co.: Singapore, 2004.

478. The Nobel Prize in Chemistry 2009. NobelPrize.org summary, accessed December 24, 2023.

479. Ogle, J. M.; Brodersen, D. E.; Clemons, W. M.; Tarry, M. J.; Carter, A. P.; Ramakrishnan, V. Recognition of Cognate Transfer RNA by the 30S Ribosomal Subunit. *Science* **2001**, *292* (5518), 897–902. https://doi.org/10.1126/science.1060612

480. Ramakrishnan, V. Unraveling the Structure of the Ribosome (Nobel Lecture). *Angew. Chem. Int. Ed.* **2010**, *49* (26), 4355–4380. https://doi.org/10.1002/anie.201001436

481. Schmeing, T. M.; Voorhees, R. M.; Kelley, A. C.; Gao, Y.-G.; Murphy, F. V.; Weir, J. R.; Ramakrishnan, V. The Crystal Structure of the Ribosome Bound to EF-Tu and Aminoacyl-tRNA. *Science* **2009**, *326* (5953), 688–694. https://doi.org/10.1126/science.1179700

482. Schuette, J.; Murphy, F. V.; Kelley, A. C.; Weir, J. R.; Giesebrecht, J.; Connell, S. R.; Loerke, J.; Mielke, T.; Zhang, W.; Penczek, P. A.; Ramakrishnan, V.; Spahn, C. M. T. GTPase Activation of Elongation Factor EF-Tu by the Ribosome during Decoding. *EMBO J.* **2009**, *28* (6), 755–765. https://doi.org/10.1038/emboj.2009.26

483. Steitz, T. A. From the Structure and Function of the Ribosome to New Antibiotics (Nobel Lecture). *Angew. Chem. Int. Ed.* **2010**, *49* (26), 4381–4398. https://doi.org/10.1002/anie.201000708

484. Villa, E.; Sengupta, J.; Trabuco, L. G.; LeBarron, J.; Baxter, W. T.; Shaikh, T. R.; Grassucci, R. A.; Nissen, P.; Ehrenberg, M.; Schulten, K.; Frank, J. Ribosome-Induced Changes in Elongation Factor Tu Conformation

Control GTP Hydrolysis. *Proc. Natl. Acad. Sci. U.S.A.* **2009**, *106* (4), 1063–1068. https://doi.org/10.1073/pnas.0811370106

485. Voorhees, R. M.; Schmeing, T. M.; Kelley, A. C.; Ramakrishnan, V. The Mechanism for Activation of GTP Hydrolysis on the Ribosome. *Science* **2010**, *330* (6005), 835–838. https://doi.org/10.1126/science.1194460
486. Agirrezabala, X.; Frank, J. Elongation in Translation as a Dynamic Interaction among the Ribosome, tRNA, and Elongation Factors EF-G and EF-Tu. *Quat. Rev. Biophys.* **2009**, *42* (3), 159–200. https://doi.org/10.1017/S0033583509990060
487. Schmeing, T. M.; Ramakrishnan, V. What Recent Ribosome Structures Have Revealed about the Mechanism of Translation. *Nature* **2009**, *461* (7268), 1234–1242. https://doi.org/10.1038/nature08403
488. Yonath, A. Hibernating Bears, Antibiotics, and the Evolving Ribosome (Nobel Lecture). *Angew. Chem. Int. Ed.* **2010**, *49* (26), 4340–4354. https://doi.org/10.1002/anie.201001297
489. Adamczyk, A. J.; Warshel, A. Converting Structural Information into an Allosteric-Energy-Based Picture for Elongation Factor Tu Activation by the Ribosome. *Proc. Natl. Acad. Sci. U.S.A.* **2011**, *108* (24), 9827–9832. https://doi.org/10.1073/pnas.1105714108
490. Constanciel, R.; Contreras, R. Self Consistent Field Theory of Solvent Effects Representation by Continuum Models: Introduction of Desolvation Contribution. *Theor. Chim. Acta* **1984**, *65* (1), 1–11. https://doi.org/10.1007/PL00020119
491. Åqvist, J.; Kamerlin, S. C. L. Conserved Motifs in Different Classes of GTPases Dictate Their Specific Modes of Catalysis. *ACS Catal.* **2016**, *6* (3), 1737–1743. https://doi.org/10.1021/acscatal.5b02491
492. Mondal, D.; Warshel, A. EF-Tu and EF-G Are Activated by Allosteric Effects. *Proc. Natl. Acad. Sci. U.S.A.* **2018**, *115* (13), 3386–3391. https://doi.org/10.1073/pnas.1800054115

493. Sondek, J.; Lambright, D. G.; Noel, J. P. GTPase Mechanism of G-proteins from the 1.7-Å Crystal Structure of Transducin A·GOP·AIF_4. *Nature* **1994**, *372*, 276–279.

494. Lambright, D. G.; Noel, J. P.; Hamm, H. E.; Sigler, P. B. Structural Determinants for Activation of the α-Subunit of a Heterotrimeric G Protein. *Nature* **1994**, *369* (6482), 621–628. https://doi.org/10.1038/369621a0

495. Bohm, A.; Gaudet, R.; Sigler, P. B. Structural Aspects of Heterotrimeric G-Protein Signaling. *Curr. Opin. Biotechnol.* **1997**, *8* (4), 480–487. https://doi.org/10.1016/S0958-1669(97)80072-9

496. Sunahara, R. K.; Tesmer, J. J. G.; Gilman, A. G.; Sprang, S. R. Crystal Structure of the Adenylyl Cyclase Activator Gsα. *Science* **1997**, *278* (5345), 1943–1947. https://doi.org/10.1126/science.278.5345.1943

497. Scrima, A.; Thomas, C.; Deaconescu, D.; Wittinghofer, A. The Rap–RapGAP Complex: GTP Hydrolysis without Catalytic Glutamine and Arginine Residues. *EMBO J.* **2008**, *27* (7), 1145–1153. https://doi.org/10.1038/emboj.2008.30

498. McCormick, F.; Wittinghofer, A. Interactions between Ras Proteins and Their Effectors. *Curr. Opin. Biotechnol.* **1996**, *7* (4), 449–456. https://doi.org/10.1016/S0958-1669(96)80123-6

499. Geyer, M.; Herrmann, C.; Wohlgemuth, S.; Wittinghofer, A.; Kalbitzer, H. R. Structure of the Ras-Binding Domain of RalGEF and Implications for Ras Binding and Signalling. *Nat. Struct. Mol. Biol.* **1997**, *4* (9), 694–699. https://doi.org/10.1038/nsb0997-694

500. Filchtinski, D.; Sharabi, O.; Rüppel, A.; Vetter, I. R.; Herrmann, C.; Shifman, J. M. What Makes Ras an Efficient Molecular Switch: A Computational, Biophysical, and Structural Study of Ras-GDP Interactions with Mutants of Raf. *J. Mol. Biol.* **2010**, *399* (3), 422–435. https://doi.org/10.1016/j.jmb.2010.03.046

501. Muegge, I.; Schweins, T.; Warshel, A. Electrostatic Contributions to Protein–Protein Binding Affinities: Application to Rap/Raf Interaction.

Proteins Struct., Funct., Bioinf. **1998**, *30* (4), 407–423. https://doi.org/10.1002/(SICI)1097-0134(19980301)30:4<407::AID-PROT8>3.0.CO;2-F

502. Broek, D.; Toda, T.; Michaeli, T.; Levin, L.; Birchmeier, C.; Zoller, M.; Powers, S.; Wigler, M. The *S. cerevisiae* CDC25 Gene Product Regulates the RAS/Adenylate Cyclase Pathway. *Cell* **1987**, *48* (5), 789–799. https://doi.org/10.1016/0092-8674(87)90076-6
503. Bowtell, D.; Fu, P.; Simon, M.; Senior, P. Identification of Murine Homologues of the *Drosophila* Son of Sevenless Gene: Potential Activators of Ras. *Proc. Natl. Acad. Sci. U.S.A.* **1992**, *89* (14), 6511–6515. https://doi.org/10.1073/pnas.89.14.6511
504. Wiesmüller, L.; Wittinghofer, F. Signal Transduction Pathways Involving RAS. *Cell. Signal.* **1994**, *6* (3), 247–267. https://doi.org/10.1016/0898-6568(94)90030-2
505. Bos, J. L.; Rehmann, H.; Wittinghofer, A. GEFs and GAPs: Critical Elements in the Control of Small G Proteins. *Cell* **2007**, *129* (5), 865–877. https://doi.org/10.1016/j.cell.2007.05.018
506. Yang, W.; Gao, Y. Q.; Cui, Q.; Ma, J.; Karplus, M. The Missing Link between Thermodynamics and Structure in F_1-ATPase. *Proc. Natl. Acad. Sci. U.S.A.* **2003**, *100* (3), 874–879. https://doi.org/10.1073/pnas.0337432100
507. Gao, Y. Q.; Yang, W.; Karplus, M. A Structure-Based Model for the Synthesis and Hydrolysis of ATP by F_1-ATPase. *Cell* **2005**, *123* (2), 195–205. https://doi.org/10.1016/j.cell.2005.10.001
508. Dittrich, M.; Hayashi, S.; Schulten, K. On the Mechanism of ATP Hydrolysis in F_1-ATPase. *Biophys. J.* **2003**, *85* (4), 2253–2266. https://doi.org/10.1016/S0006-3495(03)74650-5
509. Grigorenko, B. L.; Rogov, A. V.; Topol, I. A.; Burt, S. K.; Martinez, H. M.; Nemukhin, A. V. Mechanism of the Myosin Catalyzed Hydrolysis of ATP as Rationalized by Molecular Modeling. *Proc. Natl. Acad. Sci. U.S.A.* **2007**, *104* (17), 7057–7061. https://doi.org/10.1073/pnas.0701727104

510. Li, G.; Cui, Q. Mechanochemical Coupling in Myosin: A Theoretical Analysis with Molecular Dynamics and Combined QM/MM Reaction Path Calculations. *J. Phys. Chem. B* **2004**, *108* (10), 3342–3357. https://doi.org/10.1021/jp0371783
511. Beke-Somfai, T.; Lincoln, P.; Nordén, B. Double-Lock Ratchet Mechanism Revealing the Role of αSER-344 in FoF_1 ATP Synthase. *Proc. Natl. Acad. Sci. U.S.A.* **2011**, *108* (12), 4828–4833. https://doi.org/10.1073/pnas.1010453108
512. Akola, J.; Jones, R. O. Density Functional Calculations of ATP Systems. 2. ATP Hydrolysis at the Active Site of Actin. *J. Phys. Chem. B* **2006**, *110* (15), 8121–8129. https://doi.org/10.1021/jp054921d
513. Mukherjee, S.; Bora, R. P.; Warshel, A. Torque, Chemistry and Efficiency in Molecular Motors: A Study of the Rotary–Chemical Coupling in F_1-ATPase. *Quat. Rev. Biophys.* **2015**, *48* (4), 395–403. https://doi.org/10.1017/S0033583515000050
514. Mukherjee, S.; Warshel, A. Dissecting the Role of the γ-Subunit in the Rotary-Chemical Coupling and Torque Generation of F_1-ATPase. *Proc. Natl. Acad. Sci. U.S.A.* **2015**, *112* (9), 2746–2751. https://doi.org/10.1073/pnas.1500979112
515. Pu, J.; Karplus, M. How Subunit Coupling Produces the γ-Subunit Rotary Motion in F_1-ATPase. *Proc. Natl. Acad. Sci. U.S.A.* **2008**, *105* (4), 1192–1197. https://doi.org/10.1073/pnas.0708746105
516. Koga, N.; Takada, S. Folding-Based Molecular Simulations Reveal Mechanisms of the Rotary Motor F_1–ATPase. *Proc. Natl. Acad. Sci. U.S.A.* **2006**, *103* (14), 5367–5372. https://doi.org/10.1073/pnas.0509642103
517. Böckmann, R. A.; Grubmüller, H. Nanoseconds Molecular Dynamics Simulation of Primary Mechanical Energy Transfer Steps in F_1-ATP Synthase. *Nat. Struct. Biol.* **2002**, *9* (3), 198–202. https://doi.org/10.1038/nsb760
518. Czub, J.; Grubmüller, H. Torsional Elasticity and Energetics of F_1-ATPase. *Proc. Natl. Acad. Sci. U.S.A.* **2011**, *108* (18), 7408–7413. https://doi.org/10.1073/pnas.1018686108

519. Wang, H.; Oster, G. Energy Transduction in the F_1 Motor of ATP Synthase. *Nature* **1998**, *396* (6708), 279–282. https://doi.org/10.1038/24409

520. Hyeon, C.; Onuchic, J. N. Mechanical Control of the Directional Stepping Dynamics of the Kinesin Motor. *Proc. Natl. Acad. Sci. U.S.A.* **2007**, *104* (44), 17382–17387. https://doi.org/10.1073/pnas.0708828104

521. Hyeon, C.; Thirumalai, D. Capturing the Essence of Folding and Functions of Biomolecules Using Coarse-Grained Models. *Nat. Commun.* **2011**, *2*, 487. https://doi.org/10.1038/ncomms1481

522. Hyeon, C.; Lorimer, G. H.; Thirumalai, D. Dynamics of Allosteric Transitions in GroEL. *Proc. Natl. Acad. Sci. U.S.A.* **2006**, *103* (50), 18939–18944. https://doi.org/10.1073/pnas.0608759103

523. Falke, S.; Tama, F.; Brooks, C. L.; Gogol, E. P.; Fisher, M. T. The 13Å Structure of a Chaperonin GroEL–Protein Substrate Complex by Cryo-Electron Microscopy. *J. Mol. Biol.* **2005**, *348* (1), 219–230. https://doi.org/10.1016/j.jmb.2005.02.027

524. Furuike, S.; Hossain, M. D.; Maki, Y.; Adachi, K.; Suzuki, T.; Kohori, A.; Itoh, H.; Yoshida, M.; Kinosita, K. Axle-Less F_1-ATPase Rotates in the Correct Direction. *Science* **2008**, *319* (5865), 955–958. https://doi.org/10.1126/science.1151343

525. Chiwata, R.; Kohori, A.; Kawakami, T.; Shiroguchi, K.; Furuike, S.; Adachi, K.; Sutoh, K.; Yoshida, M.; Kinosita, K. None of the Rotor Residues of F_1-ATPase Are Essential for Torque Generation. *Biophys. J.* **2014**, *106* (10), 2166–2174. https://doi.org/10.1016/j.bpj.2014.04.013

526. Mukherjee, S.; Warshel, A. Electrostatic Origin of the Mechanochemical Rotary Mechanism and the Catalytic Dwell of F_1-ATPase. *Proc. Natl. Acad. Sci. U.S.A.* **2011**, *108* (51), 20550–20555. https://doi.org/10.1073/pnas.1117024108

527. Mukherjee, S.; Warshel, A. Realistic Simulations of the Coupling between the Protomotive Force and the Mechanical Rotation of the F°-ATPase. *Proc. Natl. Acad. Sci. U.S.A.* **2012**, *109* (37), 14876–14881. https://doi.org/10.1073/pnas.1212841109

528. Bai, C.; Warshel, A. Revisiting the Protomotive Vectorial Motion of F°-ATPase. *Proc. Natl. Acad. Sci. U.S.A.* **2019**, *116* (39), 19484–19489. https://doi.org/10.1073/pnas.1909032116
529. Astumian, R. D.; Mukherjee, S.; Warshel, A. The Physics and Physical Chemistry of Molecular Machines. *ChemPhysChem* **2016**, *17* (12), 1719–1741. https://doi.org/10.1002/cphc.201600184
530. Mukherjee, S.; Warshel, A. (Duplicate of 526): Electrostatic Origin of the Mechanochemical Rotary Mechanism and the Catalytic Dwell of F_1-ATPase. *Proc. Natl. Acad. Sci. U.S.A.* **2011**, *108* (51), 20550–20555. https://doi.org/10.1073/pnas.1117024108
531. Kamerlin, S. C. L.; Warshel, A. On the Energetics of ATP Hydrolysis in Solution. *J. Phys. Chem. B* **2009**, *113* (47), 15692–15698. https://doi.org/10.1021/jp907223t
532. Mukherjee, S.; Warshel, A. Electrostatic Origin of the Unidirectionality of Walking Myosin V Motors. *Proc. Natl. Acad. Sci. U.S.A.* **2013**, *110* (43), 17326–17331. https://doi.org/10.1073/pnas.1317641110
533. Alhadeff, R.; Warshel, A. Reexamining the Origin of the Directionality of Myosin V. *Proc. Natl. Acad. Sci. U.S.A.* **2017**, *114* (39), 10426–10431. https://doi.org/10.1073/pnas.1711214114
534. Halder, R.; Chu, Z. T.; Ti, R.; Zhu, L.; Warshel, A. On the Control of Directionality of Myosin. *J. Am. Chem. Soc.* **2024**, *146* (41), 28360–28374. https://doi.org/10.1021/jacs.4c09528
535. Rychkova, A.; Mukherjee, S.; Bora, R. P.; Warshel, A. Simulating the Pulling of Stalled Elongated Peptide from the Ribosome by the Translocon. *Proc. Natl. Acad. Sci. U.S.A.* **2013**, *110* (25), 10195–10200. https://doi.org/10.1073/pnas.1307869110
536. Rychkova, A.; Warshel, A. On the Nature of the Apparent Free Energy of Inserting Amino Acids into Membrane through the Translocon. *J. Phys. Chem. B* **2013**, *117* (44), 13748–13754. https://doi.org/10.1021/jp406925y

537. Ismail, N.; Hedman, R.; Schiller, N.; von Heijne, G. A Biphasic Pulling Force Acts on Transmembrane Helices during Translocon-Mediated Membrane Integration. *Nat. Struct. Mol. Biol.* **2012**, *19* (10), 1018–1022. https://doi.org/10.1038/nsmb.2376

538. Kaila, V. R. I.; Verkhovsky, M. I.; Wikström, M. Proton-Coupled Electron Transfer in Cytochrome Oxidase. *Chem. Rev.* **2010**, *110* (12), 7062–7081. https://doi.org/10.1021/cr1002003

539. Yoshikawa, S.; Muramoto, K.; Shinzawa-Itoh, K. Proton-Pumping Mechanism of Cytochrome c Oxidase. *Annu. Rev. Biophys.* **2011**, *40*, 205–223. https://doi.org/10.1146/annurev-biophys-042910-155341

540. Ferguson-Miller, S.; Hiser, C.; Liu, J. Gating and Regulation of the Cytochrome c Oxidase Proton Pump. *Biochim. Biophys. Acta (BBA) — Bioenergetics* **2012**, *1817* (4), 489–494. https://doi.org/10.1016/j.bbabio.2011.11.018

541. von Ballmoos, C.; Ädelroth, P.; Gennis, R. B.; Brzezinski, P. Proton Transfer in Ba$_3$ Cytochrome c Oxidase from *Thermus thermophilus*. *Biochim. Biophys. Acta (BBA) – Bioenergetics* **2012**, *1817* (4), 650–657. https://doi.org/10.1016/j.bbabio.2011.11.015

542. Popović, D. M.; Stuchebrukhov, A. A. Coupled Electron and Proton Transfer Reactions during the O→E Transition in Bovine Cytochrome c Oxidase. *Biochim. Biophys. Acta (BBA) – Bioenergetics* **2012**, *1817* (4), 506–517. https://doi.org/10.1016/j.bbabio.2011.10.013

543. Wikström, M. Active Site Intermediates in the Reduction of O_2 by Cytochrome Oxidase, and Their Derivatives. *Biochim. Biophys. Acta (BBA) – Bioenergetics* **2012**, *1817* (4), 468–475. https://doi.org/10.1016/j.bbabio.2011.10.010

544. Blomberg, M. R. A.; Siegbahn, P. E. M. The Mechanism for Proton Pumping in Cytochrome c Oxidase from an Electrostatic and Quantum Chemical Perspective. *Biochim. Biophys. Acta (BBA) – Bioenergetics* **2012**, *1817* (4), 495–505. https://doi.org/10.1016/j.bbabio.2011.09.014

545. Michel, H.; Behr, J.; Harrenga, A.; Kannt, A. Cytochrome c Oxidase: Structure and Spectroscopy. *Annu. Rev. Biophys.* **1998**, 27, 329–356. https://doi.org/10.1146/annurev.biophys.27.1.329

546. Wikström, M. Proton Translocation by Cytochrome c Oxidase: A Rejoinder to Recent Criticism. *Biochemistry* **2000**, *39* (13), 3515–3519. https://doi.org/10.1021/bi9925322

547. Brzezinski, P.; Larsson, G. Redox-Driven Proton Pumping by Heme-Copper Oxidases. *Biochim. Biophys. Acta (BBA) – Bioenergetics* **2003**, *1605* (1), 1–13. https://doi.org/10.1016/S0005-2728(03)00079-3

548. Hosler, J. P.; Ferguson-Miller, S.; Mills, D. A. Energy Transduction: Proton Transfer Through the Respiratory Complexes. *Annu. Rev. Biochem.* **2006**, *75*, 165–187. https://doi.org/10.1146/annurev.biochem.75.062003.101730

549. Yoshikawa, S.; Muramoto, K.; Shinzawa-Itoh, K.; Mochizuki, M. Structural Studies on Bovine Heart Cytochrome c Oxidase. *Biochim. Biophys. Acta (BBA) – Bioenergetics* **2012**, *1817* (4), 579–589. https://doi.org/10.1016/j.bbabio.2011.12.012

550. Sarti, P.; Arese, M.; Forte, E.; Giuffrè, A.; Mastronicola, D. Mitochondria and Nitric Oxide: Chemistry and Pathophysiology. In *Advances in Mitochondrial Medicine*; Scatena, R., Bottoni, P., Giardina, B., Eds.; Springer Netherlands: Dordrecht, **2012**; pp 75–92. https://doi.org/10.1007/978-94-007-2869-1_4

551. Kadenbach, B.; Hüttemann, M. The Subunit Composition and Function of Mammalian Cytochrome c Oxidase. *Mitochondrion* **2015**, *24*, 64–76. https://doi.org/10.1016/j.mito.2015.07.002

552. Yoshikawa, S.; Shimada, A.; Shinzawa-Itoh, K. Respiratory Conservation of Energy with Dioxygen: Cytochrome c Oxidase. In *Sustaining Life on Planet Earth: Metalloenzymes Mastering Dioxygen and Other Chewy Gases*; Kroneck, P. M. H.; Sosa Torres, M. E., Eds.; Springer International Publishing: Cham, 2015; pp 89–130. https://doi.org/10.1007/978-3-319-12415-5_4

553. Yoshikawa, S.; Shimada, A. Reaction Mechanism of Cytochrome c Oxidase. *Chem. Rev.* **2015**, *115* (4), 1936–1989. https://doi.org/10.1021/cr500266a
554. Ostermeier, C.; Harrenga, A.; Ermler, U.; Michel, H. Structure at 2.7 Å Resolution of the Paracoccus denitrificans Two-Subunit Cytochrome c Oxidase Complexed with an Antibody FV Fragment. *Proc. Natl. Acad. Sci. U.S.A.* **1997**, *94* (20), 10547–10553. https://doi.org/10.1073/pnas.94.20.10547
555. Hino, T.; Matsumoto, Y.; Nagano, S.; Sugimoto, H.; Fukumori, Y.; Murata, T.; Iwata, S.; Shiro, Y. Structural Basis of Biological N_2O Generation by Bacterial Nitric Oxide Reductase. *Science* **2010**, *330* (6011), 1666–1670. https://doi.org/10.1126/science.1195591
556. Tiefenbrunn, T.; Liu, W.; Chen, Y.; Katritch, V.; Stout, C. D.; Fee, J. A.; Cherezov, V. High-Resolution Structure of the Ba_3 Cytochrome c Oxidase from *Thermus thermophilus* in a Lipidic Environment. *PLOS ONE* **2011**, *6* (7), e22348. https://doi.org/10.1371/journal.pone.0022348
557. Han, D.; Morgan, J. E.; Gennis, R. B. G204D, a Mutation That Blocks the Proton-Conducting D-Channel of the Aa_3-Type Cytochrome c Oxidase from *Rhodobacter sphaeroides*. *Biochemistry* **2005**, *44* (38), 12767–12774. https://doi.org/10.1021/bi051141m
558. Brändén, G.; Pawate, A. S.; Gennis, R. B.; Brzezinski, P. Controlled Uncoupling and Recoupling of Proton Pumping in Cytochrome c Oxidase. *Proc. Natl. Acad. Sci. U.S.A.* **2006**, *103* (2), 317–322. https://doi.org/10.1073/pnas.0507734103
559. Belevich, I.; Bloch, D. A.; Belevich, N.; Wikström, M.; Verkhovsky, M. I. Exploring the Proton Pump Mechanism of Cytochrome c Oxidase in Real Time. *Proc. Natl. Acad. Sci. U.S.A.* **2007**, *104* (8), 2685–2690. https://doi.org/10.1073/pnas.0608794104
560. Namslauer, A.; Lepp, H.; Braöndén, M.; Jasaitis, A.; Verkhovsky, M. I.; Brzezinski, P. Plasticity of Proton Pathway Structure and Water Coordination

in Cytochrome c Oxidase. *J. Biol. Chem.* **2007**, *282* (20), 15148–15158. https://doi.org/10.1074/jbc.M700348200

561. Kannt, A.; Lancaster, C. R. D.; Michel, H. The Coupling of Electron Transfer and Proton Translocation: Electrostatic Calculations on *Paracoccus denitrificans* Cytochrome c Oxidase. *Biophys. J.* **1998**, *74* (2), 708–721. https://doi.org/10.1016/S0006-3495(98)73996-7
562. Michel, H. Cytochrome c Oxidase: Catalytic Cycle and Mechanisms of Proton Pumping – A Discussion. *Biochemistry* **1999**, *38* (46), 15129–15140. https://doi.org/10.1021/bi9910934
563. Lee, H.; Das, T. K.; Rousseau, D. L.; Mills, D.; Ferguson-Miller, S.; Gennis, R. B. Mutations in the Putative H-Channel in the Cytochrome c Oxidase from *Rhodobacter sphaeroides* Show That This Channel Is Not Important for Proton Conduction but Reveal Modulation of the Properties of Heme a. *Biochemistry* **2000**, *39* (11), 2989–2996. https://doi.org/10.1021/bi9924821
564. Cai, X.; Son, C. Y.; Mao, J.; Kaur, D.; Zhang, Y.; Khaniya, U.; Cui, Q.; Gunner, M. R. Identifying the Proton Loading Site Cluster in the Ba_3 Cytochrome c Oxidase That Loads and Traps Protons. *Biochim. Biophys. Acta (BBA) – Bioenergetics* **2020**, *1861* (10), 148239. https://doi.org/10.1016/j.bbabio.2020.148239
565. Liang, R.; Swanson, J. M. J.; Peng, Y.; Wikström, M.; Voth, G. A. Multiscale Simulations Reveal Key Features of the Proton-Pumping Mechanism in Cytochrome c Oxidase. *Proc. Natl. Acad. Sci. U.S.A.* **2016**, *113* (27), 7420–7425. https://doi.org/10.1073/pnas.1601982113
566. Salomonsson, L.; Faxén, K.; Ädelroth, P.; Brzezinski, P. The Timing of Proton Migration in Membrane-Reconstituted Cytochrome c Oxidase. *Proc. Natl. Acad. Sci. U.S.A.* **2005**, *102* (49), 17624–17629. https://doi.org/10.1073/pnas.0505431102
567. Vakkasoglu, A. S.; Morgan, J. E.; Han, D.; Pawate, A. S.; Gennis, R. B. Mutations Which Decouple the Proton Pump of the Cytochrome c

Oxidase from *Rhodobacter sphaeroides* Perturb the Environment of Glutamate 286. *FEBS Lett.* **2006**, *580* (19), 4613–4617. https://doi.org/10.1016/j.febslet.2006.07.036

568. Gorbikova, E. A.; Belevich, N. P.; Wikström, M.; Verkhovsky, M. I. Time-Resolved ATR-FTIR Spectroscopy of the Oxygen Reaction in the D124N Mutant of Cytochrome c Oxidase from *Paracoccus denitrificans*. *Biochemistry* **2007**, *46* (45), 13141–13148. https://doi.org/10.1021/bi701614w
569. Buschmann, S.; Warkentin, E.; Xie, H.; Langer, J. D.; Ermler, U.; Michel, H. The Structure of Cbb_3 Cytochrome Oxidase Provides Insights into Proton Pumping. *Science* **2010**, *329* (5989), 327–330. https://doi.org/10.1126/science.1187303
570. Lyons, J. A.; Aragão, D.; Slattery, O.; Pisliakov, A. V.; Soulimane, T.; Caffrey, M. Structural Insights into Electron Transfer in Caa_3-Type Cytochrome Oxidase. *Nature* **2012**, *487* (7408), 514–518. https://doi.org/10.1038/nature11182
571. Hematian, S.; Garcia-Bosch, I.; Karlin, K. D. Synthetic Heme/Copper Assemblies: Toward an Understanding of Cytochrome c Oxidase Interactions with Dioxygen and Nitrogen Oxides. *Acc. Chem. Res.* **2015**, *48* (8), 2462–2474. https://doi.org/10.1021/acs.accounts.5b00265
572. Sezer, M.; Kielb, P.; Kuhlmann, U.; Mohrmann, H.; Schulz, C.; Heinrich, D.; Schlesinger, R.; Heberle, J.; Weidinger, I. M. Surface-Enhanced Resonance Raman Spectroscopy Reveals Potential Induced Redox and Conformational Changes of Cytochrome c Oxidase on Electrodes. *J. Phys. Chem. B* **2015**, *119* (30), 9586–9591. https://doi.org/10.1021/acs.jpcb.5b03206
573. Yano, N.; Muramoto, K.; Mochizuki, M.; Shinzawa-Itoh, K.; Yamashita, E.; Yoshikawa, S.; Tsukihara, T. X-Ray Structure of Cyanide-Bound Bovine Heart Cytochrome c Oxidase in the Fully Oxidized State at 2.0 Å Resolution. *Acta Crystallogr. F* **2015**, *71* (6), 726–730. https://doi.org/10.1107/S2053230X15007025

574. Vilhjálmsdóttir, J.; Johansson, A. L.; Brzezinski, P. Structural Changes and Proton Transfer in Cytochrome c Oxidase. *Sci. Rep.* **2015**, *5*, 12047. https://doi.org/10.1038/srep12047

575. Zitare, U.; Alvarez-Paggi, D.; Morgada, M. N.; Abriata, L. A.; Vila, A. J.; Murgida, D. H. Reversible Switching of Redox-Active Molecular Orbitals and Electron Transfer Pathways in Cu(A) Sites of Cytochrome c Oxidase. *Angew. Chem. Int. Ed.* **2015**, *127* (33), 9691–9695. https://doi.org/10.1002/ange.201504188

576. Goyal, P.; Yang, S.; Cui, Q. Microscopic Basis for Kinetic Gating in Cytochrome c Oxidase: Insights from QM/MM Analysis. *Chem. Sci.* **2015**, 6, 826–841. https://doi.org/10.1039/C4SC01674B

577. Sharma, V.; Enkavi, G.; Vattulainen, I.; Róg, T.; Wikström, M. Proton-Coupled Electron Transfer and the Role of Water Molecules in Proton Pumping by Cytochrome c Oxidase. *Proc. Natl. Acad. Sci. U.S.A.* **2015**, *112* (7), 2040–2045. https://doi.org/10.1073/pnas.1409543112

578. Oliveira, A. S. F.; Campos, S. R. R.; Baptista, A. M.; Soares, C. M. Coupling between Protonation and Conformation in Cytochrome c Oxidase: Insights from Constant-pH MD Simulations. *Biochim. Biophys. Acta (BBA) – Bioenergetics* **2016**, *1857* (6), 759–771. https://doi.org/10.1016/j.bbabio.2016.03.024

579. Blomberg, M. R. A.; Siegbahn, P. E. M.; Warshel, A. Exploring Pathways and Barriers for Coupled ET/PT in Cytochrome c Oxidase: A General Framework for Examining Energetics and Mechanistic Alternatives. *Biochim. Biophys. Acta (BBA) – Bioenergetics* **2007**, *1767* (3), 244–260. https://doi.org/10.1016/j.bbabio.2007.01.015

580. Song, Y.; Mao, J.; Gunner, M. R. Electrostatic Environment of Hemes in Proteins: pK_as of Hydroxyl Ligands. *Biochemistry* **2006**, *45* (26), 7949–7958. https://doi.org/10.1021/bi052182l

581. Popović, D. M.; Stuchebrukhov, A. A. Electrostatic Study of the Proton Pumping Mechanism in Bovine Heart Cytochrome c Oxidase. *J. Am.*

Chem. Soc. **2004**, *126* (6), 1858–1871. https://doi.org/10.1021/ja038267w

582. Xu, J.; Voth, G. A. Computer Simulation of Explicit Proton Translocation in Cytochrome c Oxidase: The D-Pathway. *Proc. Natl. Acad. Sci. U.S.A.* **2005**, *102* (19), 6795–6800. https://doi.org/10.1073/pnas.0408117102
583. Ghosh, N.; Prat-Resina, X.; Gunner, M. R.; Cui, Q. Microscopic pK_a Analysis of Glu286 in Cytochrome c Oxidase (*Rhodobacter sphaeroides*): Toward a Calibrated Molecular Model. *Biochemistry* **2009**, *48* (11), 2468–2485. https://doi.org/10.1021/bi8021284
584. Lee, H. J.; Svahn, E.; Swanson, J. M. J.; Lepp, H.; Voth, G. A.; Brzezinski, P.; Gennis, R. B. Intricate Role of Water in Proton Transport through Cytochrome c Oxidase. *J. Am. Chem. Soc.* **2010**, *132* (45), 16225–16239. https://doi.org/10.1021/ja107244g
585. Yang, S.; Cui, Q. Glu-286 Rotation and Water Wire Reorientation Are Unlikely the Gating Elements for Proton Pumping in Cytochrome c Oxidase. *Biophys. J.* **2011**, *101* (1), 61–69. https://doi.org/10.1016/j.bpj.2011.05.004
586. Peng, Y.; Voth, G. A. Expanding the View of Proton Pumping in Cytochrome c Oxidase through Computer Simulation. *Biochim. Biophys. Acta (BBA) – Bioenergetics* **2012**, *1817* (4), 518–525. https://doi.org/10.1016/j.bbabio.2011.11.017
587. Chakrabarty, S.; Namslauer, I.; Brzezinski, P.; Warshel, A. Exploration of the Cytochrome c Oxidase Pathway Puzzle and Examination of the Origin of Elusive Mutational Effects. *Biochim. Biophys. Acta (BBA) – Bioenergetics* **2011**, *1807* (4), 413–426. https://doi.org/10.1016/j.bbabio.2011.01.004
588. Alhadeff, R.; Warshel, A. Simulating the Function of Sodium/Proton Antiporters. *Proc. Natl. Acad. Sci. U.S.A.* **2015**, *112* (40), 12378–12383. https://doi.org/10.1073/pnas.1516881112

589. Alhadeff, R.; Warshel, A. Simulating the Function of the MjNhaP1 Transporter. *J. Phys. Chem. B* **2016**, *120* (42), 10951–10958. https://doi.org/10.1021/acs.jpcb.6b08126

590. Jia, Y.; Jia, G. Role of Intestinal Na^+/H^+ Exchanger Inhibition in the Prevention of Cardiovascular and Kidney Disease. *Ann. Transl. Med.* **2015**, *3* (7), 91. https://doi.org/10.3978/j.issn.2305-5839.2015.02.26

591. Counillon, L.; Bouret, Y.; Marchiq, I.; Pouysségur, J. Na^+/H^+ Antiporter (NHE1) and Lactate/H^+ Symporters (MCTs) in pH Homeostasis and Cancer Metabolism. *Biochim. Biophys. Acta (BBA) – Mol. Cell Res.* **2016**, *1863* (10), 2465–2480. https://doi.org/10.1016/j.bbamcr.2016.02.018

592. Higgins, C. F. Multiple Molecular Mechanisms for Multidrug Resistance Transporters. *Nature* **2007**, *446* (7137), 749–757. https://doi.org/10.1038/nature05630

593. Fischbach, M. A.; Walsh, C. T. Antibiotics for Emerging Pathogens. *Science* **2009**, *325* (5944), 1089–1093. https://doi.org/10.1126/science.1176667

594. Staud, F.; Cerveny, L.; Ahmadimoghaddam, D.; Ceckova, M. Multidrug and Toxin Extrusion Proteins (MATE/SLC47); Role in Pharmacokinetics. *Int. J. Biochem. Cell Biol.* **2013**, *45* (9), 2007–2011. https://doi.org/10.1016/j.biocel.2013.06.022

595. Garrett, R. H.; Grisham, C. M. *Biochemistry*, 3rd ed.; Thomson Brooks/Cole, 2007.

596. Rasmussen, S. G. F.; Choi, H.-J.; Rosenbaum, D. M.; Kobilka, T. S.; Thian, F. S.; Edwards, P. C.; Burghammer, M.; Ratnala, V. R. P.; Sanishvili, R.; Fischetti, R. F.; Schertler, G. F. X.; Weis, W. I.; Kobilka, B. K. Crystal Structure of the Human β_2 Adrenergic G-Protein-Coupled Receptor. *Nature* **2007**, *450* (7168), 383–387. https://doi.org/10.1038/nature06325

597. Oldham, W. M.; Hamm, H. E. Structural Basis of Function in Heterotrimeric G Proteins. *Quat. Rev. Biophys.* **2006**, *39* (2), 117–166. https://doi.org/10.1017/S0033583506004306

598. Hofmann, K. P.; Scheerer, P.; Hildebrand, P. W.; Choe, H.-W.; Park, J. H.; Heck, M.; Ernst, O. P. A G Protein-Coupled Receptor at Work: The Rhodopsin Model. *Trends Biochem. Sci.* **2009**, *34* (11), 540–552. https://doi.org/10.1016/j.tibs.2009.07.005

599. Oldham, W. M.; Hamm, H. E. Structural Basis of Function in Heterotrimeric G Proteins. *Q. Rev. Biophys.* **2006**, *39* (2), 117–166. https://doi.org/10.1017/S0033583506004306.

600. Hofmann, K. P.; Scheerer, P.; Hildebrand, P. W.; Choe, H.-W.; Park, J. H.; Heck, M.; Ernst, O. P. A G Protein-Coupled Receptor at Work: The Rhodopsin Model. *Trends Biochem. Sci.* **2009**, *34* (11), 540–552. https://doi.org/10.1016/j.tibs.2009.07.005.

601. Jastrzebska, B.; Ringler, P.; Lodowski, D. T.; Moiseenkova-Bell, V.; Golczak, M.; Müller, S. A.; Palczewski, K.; Engel, A. Rhodopsin–Transducin Heteropentamer: Three-Dimensional Structure and Biochemical Characterization. *J. Struct. Biol.* **2011**, *176* (3), 387–394. https://doi.org/10.1016/j.jsb.2011.08.016.

602. Hanson, M. A.; Roth, C. B.; Jo, E.; Griffith, M. T.; Scott, F. L.; Reinhart, G.; Desale, H.; Clemons, B.; Cahalan, S. M.; Schuerer, S. C.; Sanna, M. G.; Han, G. W.; Kuhn, P.; Rosen, H.; Stevens, R. C. Crystal Structure of a Lipid G Protein–Coupled Receptor. *Science* **2012**, *335* (6070), 851–855. https://doi.org/10.1126/science.1215904.

603. Katritch, V.; Cherezov, V.; Stevens, R. C. Diversity and Modularity of G Protein-Coupled Receptor Structures. *Trends Pharmacol. Sci.* **2012**, *33* (1), 17–27. https://doi.org/10.1016/j.tips.2011.09.003.

604. Latorraca, N. R.; Wang, J. K.; Bauer, B.; Townshend, R. J. L.; Hollingsworth, S. A.; Olivieri, J. E.; Xu, H. E.; Sommer, M. E.; Dror, R. O. Molecular Mechanism of GPCR-Mediated Arrestin Activation. *Nature* **2018**, *557* (7705), 452–456. https://doi.org/10.1038/s41586-018-0077-3.

605. Dror, R. O.; Arlow, D. H.; Maragakis, P.; Mildorf, T. J.; Pan, A. C.; Xu, H.; Borhani, D. W.; Shaw, D. E. Activation Mechanism of the β_2-Adrenergic

Receptor. *Proc. Natl. Acad. Sci. U.S.A.* **2011**, *108* (46), 18684–18689. https://doi.org/10.1073/pnas.1110499108.

606. Kohlhoff, K. J.; Shukla, D.; Lawrenz, M.; Bowman, G. R.; Konerding, D. E.; Belov, D.; Altman, R. B.; Pande, V. S. Cloud-Based Simulations on Google Exacycle Reveal Ligand Modulation of GPCR Activation Pathways. *Nat. Chem.* **2014**, *6* (1), 15–21. https://doi.org/10.1038/nchem.1821.
607. Alhadeff, R.; Vorobyov, I.; Yoon, H. W.; Warshel, A. Exploring the Free-Energy Landscape of C Activation. *Proceedings of the National Academy of Sciences* **2018**, *115* (41), 10327–10332. https://doi.org/10.1073/pnas.1810316115.
608. Alhadeff, R.; Warshel, A. A Free-Energy Landscape for the Glucagon-like Peptide 1 Receptor GLP1R. *Proteins* **2020**, *88* (1), 127–134. https://doi.org/10.1002/prot.25777.
609. Mondal, D.; Kolev, V.; Warshel, A. Exploring the Activation Pathway and Gi-Coupling Specificity of the μ-Opioid Receptor. *Proceedings of the National Academy of Sciences* **2020**, *117* (42), 26218–26225. https://doi.org/10.1073/pnas.2013364117.
610. Ti, R.; Pang, B.; Yu, L.; Gan, B.; Ma, W.; Warshel, A.; Ren, R.; Zhu, L. Fine-Tuning Activation Specificity of G-Protein-Coupled Receptors via Automated Path Searching. *Proceedings of the National Academy of Sciences* **2024**, *121* (8), e2317893121. https://doi.org/10.1073/pnas.2317893121.
611. Deupi, X.; Kobilka, B. K. Energy Landscapes as a Tool to Integrate GPCR Structure, Dynamics, and Function. *Physiology* **2010**, *25* (5), 293–303. https://doi.org/10.1152/physiol.00002.2010.
612. Bai, C.; Warshel, A. Critical Differences Between the Binding Features of the Spike Proteins of SARS-CoV-2 and SARS-CoV. *J. Phys. Chem. B* **2020**, *124* (28), 5907–5912. https://doi.org/10.1021/acs.jpcb.0c04317.
613. Mondal, D.; Warshel, A. Exploring the Mechanism of Covalent Inhibition: Simulating the Binding Free Energy of α-Ketoamide Inhibitors of

the Main Protease of SARS-CoV-2. *Biochemistry* **2020**, *59* (48), 4601–4608. https://doi.org/10.1021/acs.biochem.0c00782.

614. Davey, C. A.; Sargent, D. F.; Luger, K.; Maeder, A. W.; Richmond, T. J. Solvent Mediated Interactions in the Structure of the Nucleosome Core Particle at 1.9Å Resolution. *Journal of Molecular Biology* **2002**, *319* (5), 1097–1113. https://doi.org/10.1016/S0022-2836(02)00386-8.
615. Portillo-Ledesma, S.; Li, Z.; Schlick, T. Genome Modeling: From Chromatin Fibers to Genes. *Current Opinion in Structural Biology* **2023**, *78*, 102506. https://doi.org/10.1016/j.sbi.2022.102506.
616. Alberts, B.; Johnson, A.; Lewis, J.; Raff, M.; Roberts, K.; Walter, P. *Molecular Biology of the Cell*; Garland Science: New York, 2015.
617. Jiang, Y.; Ruta, V.; Chen, J.; Lee, A.; MacKinnon, R. The Principle of Gating Charge Movement in a Voltage-Dependent K+ Channel. *Nature* **2003**, *423* (6935), 42–48. https://doi.org/10.1038/nature01581.
618. Schmidt, D.; Jiang, Q.-X.; MacKinnon, R. Phospholipids and the Origin of Cationic Gating Charges in Voltage Sensors. *Nature* **2006**, *444* (7120), 775–779. https://doi.org/10.1038/nature05416.
619. Swartz, K. J. Sensing Voltage across Lipid Membranes. *Nature* **2008**, *456* (7224), 891–897. https://doi.org/10.1038/nature07620.
620. Hessa, T.; Kim, H.; Bihlmaier, K.; Lundin, C.; Boekel, J.; Andersson, H.; Nilsson, I.; White, S. H.; von Heijne, G. Recognition of Transmembrane Helices by the Endoplasmic Reticulum Translocon. *Nature* **2005**, *433* (7024), 377–381. https://doi.org/10.1038/nature03216.
621. Kamerlin, S. C. L.; Vicatos, S.; Dryga, A.; Warshel, A. Coarse-Grained (Multiscale) Simulations in Studies of Biophysical and Chemical Systems. *Annual Review of Physical Chemistry* **2011**, *62*, 41–64. https://doi.org/10.1146/annurev-physchem-032210-103335.
622. Wu, C.; Shea, J.-E. Coarse-Grained Models for Protein Aggregation. *Current Opinion in Structural Biology* **2011**, *21* (2), 209–220. https://doi.org/10.1016/j.sbi.2011.02.002.

623. Tozzini, V. Coarse-Grained Models for Proteins. *Current Opinion in Structural Biology* **2005**, *15* (2), 144–150. https://doi.org/10.1016/j.sbi.2005.02.005.
624. Guardiani, C.; Livi, R.; Cecconi, F. Coarse Grained Modeling and Approaches to Protein Folding. *Current Bioinformatics* **2010**, *5* (3), 217–240.
625. Hills, R. D.; Brooks, C. L. Insights from Coarse-Grained Gō Models for Protein Folding and Dynamics. *International Journal of Molecular Sciences* **2009**, *10* (3), 889–905. https://doi.org/10.3390/ijms10030889.
626. Monticelli, L.; Kandasamy, S. K.; Periole, X.; Larson, R. G.; Tieleman, D. P.; Marrink, S.-J. The MARTINI Coarse-Grained Force Field: Extension to Proteins. *Journal of Chemical Theory and Computation* **2008**, *4* (5), 819–834. https://doi.org/10.1021/ct700324x.
627. Marrink, S. J.; de Vries, A. H.; Mark, A. E. Coarse Grained Model for Semiquantitative Lipid Simulations. *Journal of Physical Chemistry B* **2004**, *108* (2), 750–760. https://doi.org/10.1021/jp036508g.
628. Marrink, S. J.; de Vries, A. H.; Tieleman, D. P. Lipids on the Move: Simulations of Mebrane Pores, Domains, Stalks and Curves. *Biochimica et Biophysica Acta (BBA) - Biomembranes* **2009**, *1788* (1), 149–168. https://doi.org/10.1016/j.bbamem.2008.10.006.
629. Marrink, S. J.; Tieleman, D. P. Perspective on the Martini Model. *Chemical Society Reviews* **2013**, *42* (16), 6801–6822. https://doi.org/10.1039/C3CS60093A.
630. Kessel, A.; Shental-Bechor, D.; Haliloglu, T.; Ben-Tal, N. Interactions of Hydrophobic Peptides with Lipid Bilayers: Monte Carlo Simulations with M2δ. *Biophysical Journal* **2003**, *85* (6), 3431–3444. https://doi.org/10.1016/S0006-3495(03)74765-1.
631. Maddox, M. W.; Longo, M. L. A Monte Carlo Study of Peptide Insertion into Lipid Bilayers: Equilibrium Conformations and Insertion Mechanisms. *Biophysical Journal* **2002**, *82* (1), 244–263. https://doi.org/10.1016/S0006-3495(02)75391-5.

632. Milik, M.; Skolnick, J. Insertion of Peptide Chains into Lipid Membranes: An off-Lattice Monte Carlo Dynamics Model. *Proteins: Structure, Function, and Bioinformatics* **1993**, *15* (1), 10–25. https://doi.org/10.1002/prot.340150104.

633. Baumgärtner, A. Insertion and Hairpin Formation of Membrane Proteins: A Mont 6Carlo Study. *Biophysical Journal* **1996**, *71* (3), 1248–1255. https://doi.org/10.1016/S0006-3495(96)79324-4.

634. Bowie, J. U. Solving the Membrane Protein Folding Problem. *Nature* **2005**, *438* (7068), 581–589. https://doi.org/10.1038/nature04395.

635. Ben-Tal, N.; Ben-Shaul, A.; Nicholls, A.; Honig, B. Free-Energy Determinants of Alpha-Helix Insertion into Lipid Bilayers. *Biophysical Journal* **1996**, *70* (4), 1803–1812. https://doi.org/10.1016/S0006-3495(96)79744-8.

636. Moll, T. S.; Thompson, T. E. Semisynthetic Proteins: Model Systems for the Study of the Insertion of Hydrophobic Peptides into Preformed Lipid Bilayers. *Biochemistry* **1994**, *33* (51), 15469–15482. https://doi.org/10.1021/bi00255a029.

637. Ulmschneider, J. P.; Andersson, M.; Ulmschneider, M. B. Determining Peptide Partitioning Properties via Computer Simulation. *The Journal of Membrane Biology* **2011**, *239* (1), 15–26. https://doi.org/10.1007/s00232-010-9324-8.

638. Kessel, A.; Haliloglu, T.; Ben-Tal, N. Interactions of the M2δ Segment of the Acetylcholine Receptor with Lipid Bilayers: A Continuum-Solvent Model Study. *Biophysical Journal* **2003**, *85* (6), 3687–3695. https://doi.org/10.1016/S0006-3495(03)74785-7.

639. Hessa, T.; White, S. H.; von Heijne, G. Membrane Insertion of a Potassium-Channel Voltage Sensor. *Science* **2005**, *307* (5714), 1427–1427. https://doi.org/10.1126/science.1109176.

640. Dorairaj, S.; Allen, T. W. On the Thermodynamic Stability of a Charged Arginine Side Chain in a Transmembrane Helix. *Proceedings of the National Academy of Sciences* **2007**, *104* (12), 4943–4948. https://doi.org/10.1073/pnas.0610470104.

641. White, S. H.; von Heijne, G. Transmembrane Helices before, during, and after Insertion. *Current Opinion in Structural Biology* **2005**, *15* (4), 378–386. https://doi.org/10.1016/j.sbi.2005.07.004.

642. Schow, E. V.; Freites, J. A.; Myint, P. C.; Bernsel, A.; von Heijne, G.; White, S. H.; Tobias, D. J. Arginine in Membranes: The Connection Between Molecular Dynamics Simulations and Translocon-Mediated Insertion Experiments. *The Journal of Membrane Biology* **2011**, *239* (1), 35–48. https://doi.org/10.1007/s00232-010-9330-x.

643. Choe, S.; Hecht, K. A.; Grabe, M. A Continuum Method for Determining Membrane Pr47tein Insertion Energies and the Problem of Charged Residues. *Journal of General Physiology* **2008**, *131* (6), 563–573. https://doi.org/10.1085/jgp.200809959.

644. Hristova, K.; Wimley, W. C. A Look at Arginine in Membranes. *The Journal of Membrane Biology* **2011**, *239* (1), 49–56. https://doi.org/10.1007/s00232-010-9323-9.

645. Li, L.; Vorobyov, I.; Allen, T. W. The Different Interactions of Lysine and Arginine Side Chains with Lipid Membranes. *J. Phys. Chem. B* **2013**, *117* (40), 11906–11920. https://doi.org/10.1021/jp405418y.

646. Johansson, A. C. V.; Lindahl, E. The Role of Lipid Composition for Insertion and Stabilization of Amino Acids in Membranes. *The Journal of Chemical Physics* **2009**, *130* (18), 185101. https://doi.org/10.1063/1.3129863.

647. Johansson, A. C. V.; Lindahl, E. Titratable Amino Acid Solvation in Lipid Membranes as a Function of Protonation State. *J. Phys. Chem. B* **2009**, *113* (1), 245–253. https://doi.org/10.1021/jp8048873.

648. Li, L.; Vorobyov, I.; Allen, T. W. Potential of Mean Force and pKa Profile Calculation for aLipid Membrane-Exposed Arginine Side Chain. *J. Phys. Chem. B* **2008**, *112* (32), 9574–9587. https://doi.org/10.1021/jp7114912.

649. MacCallum, J. L.; Bennett, W. F. D.; Tieleman, D. P. Partitioning of Amino Acid Side Chains into Lipid Bilayers: Results from Computer Simulations

and Comparison to Experiment. *Journal of General Physiology* **2007**, *129* (5), 371–377. https://doi.org/10.1085/jgp.200709745.

650. MacCallum, J. L.; Bennett, W. F. D.; Tieleman, D. P. Distribution of Amino Acids in a Lipid Bilayer from Computer Simulations. *Biophysical Journal* **2008**, *94* (9), 3393–3404.

651. Vorobyov, I.; Kim, I.; Chu, Z. T.; Warshel, A. Refining the Treatment of Membrane Proteins by Coarse-Grained Models. *Proteins* **2016**, *84* (1), 92–117. https://doi.org/10.1002/prot.24958.

652. Warshel, A. *From Kibbutz Fishponds to The Nobel Prize: Taking Molecular Functions into Cyberspace*; World Scientific Publishing Co.: Singapore, 2021. https://doi.org/10.1142/12412.

653. Olsson, M. H. M.; Sharma, P. K.; Warshel, A. Simulating Redox Coupled Proton Transfer in Cytochrome c Oxidase: Looking for the Proton Bottleneck. *FEBS Lett.* **2005**, *579* (10), 2026–2034. https://doi.org/10.1016/j.febslet.2005.02.051.

654. Clarke, T. L.; Mostoslavsky, R. DNA Repair as a Shared Hallmark in Cancer and Ageing. *Molecular Oncology* **2022**, *16*, 3352–3379. https://doi.org/10.1002/1878-0261.13285

655. Warshel, A.; Chu, Z. T. Calculations of Solvation Free Energies in Chemistry and Biology. In *Structure and Reactivity in Aqueous Solution*; Cramer, C. J., Truhlar, D. G., Eds.; ACS Symposium Series 568; American Chemical Society: Washington, DC, 1994; Chapter 6, pp. 71–94. https://doi.org/10.1021/bk-1994-0568.ch006.

Index

www.ingramcontent.com/pod-product-compliance
Lightning Source LLC
Chambersburg PA
CBHW070744281225
37357CB00004B/328
9789819820603